W9-BKK-834

Once Upon a Time We Ate Animals

Once Upon a Time We Ate Animals

The Future of Food

Roanne van Voorst

Translation by Scott Emblen-Jarrett

HarperOne
An Imprint of HarperCollins*Publishers*

HarperCollins books may be purchased for educational, business, or sales promotional use. For information, please email the Special Markets Department at SPsales@harpercollins.com.

Originally published as *Ooit aten we dieren* in Holland in 2019 by Uitgeverij Podium. English translation by Scott Emblen-Jarrett.

FIRST HARPERONE EDITION PUBLISHED 2021

Designed by Leah Carlson-Stanisic

Library of Congress Cataloging-in-Publication Data has been applied for.

ISBN 978-0-06-300588-4

21 22 23 24 25 FRI 10 9 8 7 6 5 4 3 2 1

For Lisette, who knew all along what I still had to learn to see, and for Fedde, who will see much more in his lifetime than I can possibly imagine

The future is already here—it's just not very evenly distributed.

William Gibson

The difficulty lies, not in the new ideas, but in escaping from the old ones.

John Maynard Keynes

Contents

Contents

Introduction

Inventing a New Color

Three centuries ago, the Enlightenment took place and burning people as witches or condemning them for their beliefs became illegal. Over 150 years ago (around eight generations, by my reckoning), slavery was abolished all over the world, and it became illegal to brand other human beings, hold them against their will, misuse, or mistreat them in any other way. Around 100 years ago (five generations), women in the Western democracies were given the right to vote and formally became equal to men. You and I live in similarly turbulent, important, and exciting times.

Our age will go down in history for the immense social, economic, and cultural changes that are unfolding *right now*, at the very moment that you are reading this book. It is a transformation that is taking place all over the world in all kinds of places, and it won't be long before it suddenly strikes you. You will see it in the things you buy, the work you do, the way you raise your children—even in the way you think and feel. And when it gets to that stage, it will almost be as if things were never any different.

You and I are part of the generation that will see unnecessary animal suffering become a thing of the past in large areas of the world. This means that while the consumption and use of meat or other animal products may still go on in the near future, it will become more difficult and much more expensive. It will be a choice that differs from the norm; a choice rejected by most people. Whether you support this huge change or are against it, it is already happening; you cannot halt it, for we are in the midst of it.

Try to imagine yourself standing on a green hilltop and trying to move an enormous, heavy boulder. You push against the boulder with both hands, dig your heels into the grass, tense your core and your legs . . . It doesn't work, the boulder stays where it is, and you, you swear, you puff, you groan, you think it might be impossible to budge it—and then suddenly it starts to roll. You've managed to find the boulder's tipping point. It rolls slowly at first, then faster and faster until the boulder rolls so fast that you know it will be impossible to stop even if someone tried. We have now reached the same sort of tipping point. We are all on the edge of this shift from a slow-paced to a fast-paced movement, a movement that can no longer be stopped.

We're on a Roll

Veganism is one of the fastest-growing movements in the world. More and more scientists and futurists predict that eating meat and dairy will become much less common and may even become a social taboo in the near future. A growing number of people are saying that veganism is one of the last remaining options we have to combat climate change, and their message is being heard. In the 1990s, there were about a million people worldwide who ate no meat or dairy or did not use any animal products—often because they were saddened by the plight of animals, sometimes because they believed it was bad for the environment or their body. By 2015, this number had grown at least a hundredfold—according to some, even up to around 750 million.

In 2008, the city of Ghent in Belgium was the first city in Europe to promote a weekly meat-free day in schools and other public institutions. This idea had already gained some traction in the US; the next city to implement this was in the UK, and by 2019, forty cities worldwide were taking part in the practice, and this number continues to grow.

In 2018, Australia—the country that at the start of the twenty-first century consumed the most meat in the world—had one of the fastest-growing vegan markets on the planet. More and more Aussies are choosing soy over steak. The country came in third, behind the United Arab Emirates and China, in the number of people who are choosing to eat vegan.

In the United States, not only have sales of meat alternatives (such as soy burgers and plant-based "meat" pieces that have the taste and texture reminiscent of chicken breast) increased enormously in recent years, but sales of dairy alternatives such as coconut yogurt and almond milk have also risen. By 2021, these alternatives will make up 40 percent of all milk-style drinks, compared to 25 percent in 2016. Sales of cow's milk meanwhile dropped. Dairy Farmers of America, the largest milk cooperative in the United States and the supplier of 30 percent of the country's milk, made a billion dollars less in 2018 than the year before. This trend isn't solely limited to the US but has also been noted in the Netherlands, the UK, Germany, Australia, Italy, and Canada. In January 2019, the Canadian Food Inspection Agency published new national health guidelines that recommend to "go light" on animal proteins. What do they recommend instead for a balanced diet? Plant-based, protein-rich foods.

The global egg industry is also beginning to notice the sudden sharp drop in demand for animal products: Cal-Maine Foods, a huge American egg producer, recently reported its first annual loss in more than ten years. Shares plummeted; the company CEO said that the losses were due to the increasing popularity of egg alternatives.

In light of this, smart businesspeople would do better by investing in the vegan food industry—for example, in products such as "nut cheese." By 2024, its estimated global market value will be close to $4 billion, with annual growth of about 8 percent. Or they should invest in milk-like alternatives made from oats, soy, rice, or almonds. After ninety-two years, Elmhurst, one of the oldest dairy producers in the eastern United States, recently made the decision to switch to producing only plant-based milk alternatives; the company's CEO says this is the best way to prevent future losses.

Plant-based alternatives to meat are also doing well: so well, in fact, that traditional meat producers have decided to invest en masse, often buying out vegan companies. For example, Tyson Foods, the biggest meat producer in the US, has already invested in the most popular meat alternative in the American market: Beyond Meat. Canada's largest meat distributor, Maple Leaf Foods, bought the popular plant-based brands Field Roast and Lightlife Foods. Nestlé, the biggest food and beverage company in the world, took over Sweet Earth Foods, a company that makes exclusively plant-based products (and was set up by a former CEO of Burger King). Danone took over plant-based pioneer WhiteWave, while Unilever bought the Vegetarian Butcher.

The Dutch newspaper *De Volkskrant* considered this million-dollar sale as symbolic of the "rise of meat substitutes" and noted that multinationals and "even meat producers" are currently starting to get in on the vegetarian market. International business magazine *Forbes* did not hesitate in advising investors to ride the plant-based wave, with a headline reading "Here's why you should turn your business vegan."

The World Turned Upside Down

The business world is not the only thing that has experienced an enormous transformation in recent years, with significant changes also taking place at an individual level. Recently, 39 percent of Americans have consciously decided to eat less meat or become flexitarian, primarily because they believe it is better for their health. They have shifted from traditional pork and beef products to Beyond Sausage, a meat alternative with a consistency similar to that of pork but with less fat and sodium, as well as more protein than in actual meat; or they have shifted to the Beyond Burger, another of the company's products, which counts Bill Gates, Leonardo DiCaprio, Twitter founders Biz Stone and Evan Williams, and meat-processing giant Tyson Foods among its shareholders.

In Germany, a country known for its love of sausages, 41 percent of consumers ate less meat (and more meat alternatives) in 2018 than in previous years. That same year, Dutch people spent €80 million on meat alternatives; ten years earlier, this figure was €62 million. Researchers predict that over the next few years Dutch consumers will opt more and more for plant-based food alternatives.

It certainly looks this way, as most people who have a partial or completely plant-based diet are young—and in the years to come they will be the ones in charge of buying groceries. In 2017, 42 percent of all vegans in the UK were between the ages of fifteen and thirty-four; in Australia most of them belong to the "millennial" generation, and in other countries, too, the vast majority of vegetarians and vegans belong to the new generation of consumers. Children and teenagers increasingly opt for a plant-based diet because they have concerns about the climate, disagree with the way animals are used to make food, or simply because they like the taste of meat alternatives and dairy alternatives. A plant-based lifestyle is currently not the social norm, since the number of older consumers who eat animal products is still larger than the younger demographic who have a plant-based diet. This visible shift, however, suggests that the growth of the vegan market will be explosive in the future.

In this book I will show you the lifestyle changes you can expect in the years to come. I will show you the world of the future. Not the distant future: a future that you and your children (if you have or want any) will live to see. This future will be very different in many ways from the world you grew up in and live in now. Much sooner than you think we will eat differently, work differently, use different things, take different school trips, and pamper different pets. Above all, we will think differently about what is good and what is evil.

Sigh

When we look back on all these changes in our old age, I suspect that, in retrospect, we will heave a sigh about how long it took for us to make the switch. Too long in fact. For a long time we knew that the way society treats animals and the planet was not OK. We saw documentaries about it, or videos online, or we read about it in books and newspapers, but most of us did nothing with the information.

I am guilty of this too. I became a vegetarian when I was sixteen; I stopped eating meat, but I kept eating dairy and eggs and using leather and other animal products. I didn't want to eat meat anymore because I loved animals and didn't want them to be killed just because I happened to like how they taste, and also, in all honesty, I wanted to be "different" from my friends at school. My vegetarianism was both a way of shaping my own identity and a kind of self-conceived charity work: some people cared for the elderly, while I switched from a late-night kebab to a grilled cheese sandwich. I considered this a supreme sacrifice, a good deed that exempted me from thinking more profoundly about the complex things relating to our food system. Or maybe I was just too young to be aware that simply cutting meat out of my diet would not solve all the problems that I have written about so fervently. I have no recollection of ever asking myself how the cheese on my sandwich or the mayonnaise I dipped my fries in was made, or what my new pair of cowboy boots, with the fabulous pointed toes and a perfect slightly worn look, were made of.

It took me more than fifteen years to actually start asking these questions. I was in my thirties when I read an article about dairy cattle farming. It was a Sunday afternoon, and I was sitting in my favorite coffee spot in Philadelphia, a place that only sold "good" coffee and "ethically produced milk." I enjoyed my cappuccino, thought about what I would cook for my husband that evening, and thumbed through an issue of the *New York Times*. While doing so, I stumbled across an article reporting that all young bulls born on dairy cattle farms were slaughtered right away, as they served no purpose. Standard procedure: "Bulls don't give any milk," I read, "and so they are a waste product in the production of milk." Later I learned that male chicks suffer the same fate: as soon as their sex has been identified, they are minced alive or gassed, as they can't lay eggs and are considered a "waste product" of the egg industry.

I must have come across such information before this point. The article's message was not new; the information was not classified as "news" either. It was buried at the end of a thick weekend supplement, in an article about the American dairy industry's annual investment figures. The "waste product" comment was barely an afterthought. I remember folding the newspaper and staring long and hard at the tiny bubbles in the foam of my cappuccino. I also remember feeling confused. Surely what was written in the article couldn't be true. Could it? By buying this "good" cappuccino with the "ethically produced" foam, had I indirectly contributed to the slaughter of a perfectly healthy young bull? Why, for that matter, had I not been aware of this during all my years as a vegetarian? What kind of a sick system was this that labeled perfectly healthy animals as *waste*?

That afternoon was the start of my own lengthy personal investigation into the topic of veganism: a scientific investigation of the animal product economy, but also a personal search for my own role in it.

In this book I will tell you about my quest. Not because I want to lecture you: I have wondered for a while now if my current (vegan) diet and lifestyle is "better" than the way I lived before. Clothing made without animal products is not always more environmentally friendly than its animal-based equivalents, to name but one dilemma I face. I also find it very difficult to have to turn down a dish that has been lovingly made but that is not vegan friendly. At moments like these I find myself torn between the desire to be polite and pleasant and "normal" and my choice to no longer contribute to a system that I do not support; and no matter what decision I make, I will end up feeling bad.

I am also sharing these personal struggles of mine not because my story is so important or special, but precisely because my story is *not* that special. I expect my story might be very similar to your own, whether it's a process you've already gone through, are going through now, or have yet to go through. If you recognize yourself in my story, then the research that I have done might help you understand why you have made decisions that have supported a brutal system, even while you consider yourself a kind-hearted person. Just like I consider myself to be.

Paradox

It is perhaps the greatest paradox of being human: simply because of our humanity, we often behave inhumanely. Most people consider it a scary fact that sea levels are set to rise and cause deadly flooding in other countries as a result of our food choices. Yet this is what is happening right now.

We also reel at the idea that animals suffer unimaginably because we want to use their meat, milk, eggs, or skin. But this happens too. Relatively recently a spokesperson for the United Nations called the way that we breed, keep, and slaughter farm animals "torture," and I reasonably presume you, just like I, are against torture. We would never, personally, shove a shock rod into the nose or anus of an animal; we would never twist a cow's tail if we knew it causes a huge amount of pain; we would never castrate a male piglet without anesthetic; and we would never breed chickens that are so big they can barely walk. We would never slice up, gas, or shoot healthy animals. But this is what we do on an almost daily basis by financially supporting the meat and dairy industry.

The idea that *every week* more animals are killed for human consumption than humans who have died in *all wars in human history* combined is something we can hardly imagine. It is also something we don't want to imagine at all. It sounds so . . . absurd, doesn't it? Every time I read and reread the sentence above, I immediately feel this urge to rid myself of this image, to quickly move on to the following paragraphs, to the next section of this introduction where things get a bit more pleasant again. But what it says is true: every week we kill more animals than there have been humans killed in every war in human history.

According to researchers, 108 million people died as the result of war in the twentieth century (including both world wars). Estimates of the number of people killed in wars during the whole of human history vary between 150 million and 1 billion.

Exact numbers of the number of animals we slaughter also vary wildly, but the most conventional statistics I can find—those published by the dairy and meat industries—put the total number of farm animals slaughtered each year at 66 billion. This is just the number of cows, pigs, and other farm animals, and does not include the number of fish we catch for food. Figures for fish and other aquatic animals are estimated at about 150 billion a year. If we count up all the animals we like to eat in vast amounts—fish, chickens, pigs, cows, goats, sheep—then we reach a figure of 150 million *per day*. These statistics, however, do not include the millions of animals that are killed in laboratories each year, or those that are killed for their fur, or the male chicks and young bulls that are killed right after they are born (because "waste products" are not included in these statistics). This also does not include animals who die in rodeos and bullfights each year, or racing horses and dogs that are put down after races, or animals that die young in zoos and aquariums because they are kept in captivity, or because they are considered "surplus."

If you let these facts sink in, you will probably feel the same emotions I feel every time I do so: pity, disbelief, disgust, shame. This ability to feel compassion makes humans civilized; many believe that this ability is what separates us from animals.

I would also argue that this ability to feel compassion makes our behavior equally *uncivilized* at times. We turn a blind eye to cruelty, not because we don't care, but precisely because our deep human values are inconsistent with how we treat animals in our time. The information we are fed about this, which comes to us via newspaper articles, shocking video images that appear on social media, and now via the words written on these pages, makes us so uncomfortable that we can do nothing other than immediately distance ourselves from it. We ignore it; we act as if it's not happening. I fear this was the reason I previously ignored dozens of articles about the dairy industry, before the news truly sunk in, that afternoon in a café in Philadelphia. It was too much, it was too terrible, it seemed illogical that we—as intelligent, decent, caring people—could do this.

Tacit Evil

Yet we still do it. Historian Yuval Noah Harari wrote in the *Guardian* in 2015 that the way we treat industrially farmed animals is one of the greatest crimes in human history. With this statement I don't think he wants to deny that terrible crimes have been committed against humanity, and it's also not useful or appropriate to compare the fates of Holocaust or other genocide victims with those of animals that have fallen victim to our lifestyle: it's not a competition of suffering, after all. His statement does point to a shocking conclusion, however: most of us are sponsors of criminal activities, whether we are aware of it or not. While it is true that most of us do not harm any animals personally, we do pay others to do it for us. We do it every time we buy a box of eggs, or a cup of yogurt, or a steak. We do it every time we see an article or video that exposes the abuses in the meat and dairy industries and simply turn the page or click away. We don't *mean* any harm by this, and we also don't *feel* like we are causing harm when we do this. We justify this denial with the idea that there is no other way, that this is simply how the world works.

Yet if we avoid thinking about bad things, this results in a tacit acceptance of what goes on, and this is dangerous if this is the norm, as it leads to enormous suffering. Seen in this light, our silent generation is guilty of massive-scale animal abuse that takes place every second of every day, and we are equally guilty of the destruction of the planet, caused by the emissions created by industrial livestock farming.

We are guilty of it right now.

And now.

And now.

Albert Einstein said, "The world is in greater peril from

those who tolerate or encourage evil than from those who actually commit it." The philosopher Hannah Arendt added to this sentiment by stating that the greatest evil in the world is done by people who don't necessarily think about whether what they are doing is bad, but who simply go along with what others do and with what the norm is.

Before all this moralizing starts putting you off, don't worry: this book goes much further than simply listing accusations about the way our society treats animals and the planet. Others have done this before me in the form of books, articles, reports, and documentaries, and I feel that repeating what they have said will not help you or me think about the way we live and consume (if you are interested, however, you will find a list of reliable sources at the back of this book).

We aren't lacking knowledge on the subject; if we are open to the idea, all the information has been available for free for a long time now. We also clearly don't lack the ability to empathize; you may have noticed it just now, when I persuaded you to take part in my thought experiment about war victims and slaughtered animals.

I believe what we are lacking instead is a clear idea of what the alternatives might look like. I am a futures anthropologist by profession, meaning that I obtained a PhD in anthropology (in 2014) and was also trained in future foresighting. My research focus has varied over the past years, but it always entails the building of future scenarios and a thought experiment about how a certain scenario may impact our society, daily lives, behavior, and feelings. From my point of view, what we are currently missing is a serious exploration of a future world in which most of us live a plant-based lifestyle, where we no longer use animals for food, clothing, or other items.

Learning to See New Colors

It is not that strange if you can't imagine something like this to be possible, let alone what such a world might look like. Try to imagine an entirely new color, or a new taste, or a new smell. Something that no one has ever seen, smelled, or tasted!

Give up? Don't beat yourself up over it; it's virtually impossible.

The most our brain can do is try to mix colors or smells or tastes that we already know, which may give us an idea of a combination that has never been made before. A new combination, however, is not something completely new.

We also encounter the same problem when it comes to a plant-based lifestyle, and I believe this is the second reason why I must have come across this information without having it affect me. I had to shrug my shoulders because I couldn't think of a solution for it: That's the way things have always worked, that's just the way it is, right?

For a long time it was, yes. Just like me, you have been brought up in an age where using and eating animals are seen as completely normal. They are part of our clothes, our shoes, our candles, our apple juice, even our condoms! We grew up in an age where our parents, doctors, teachers taught us that meat and dairy were not only good for our health but were *necessary* for a healthy lifestyle. Research has proven this wrong, but we'll come back to that later. It is difficult, however, to suddenly stop believing something that you have believed for so long, especially if you are lacking an alternative vision of how you can still eat good food and stay fit and healthy.

This book will outline that vision for you. I will construct for you a futuristic dreamscape of a more animal- and environmentally friendly world with new colors, tastes, and smells. I can tell you in concrete terms what we can and will do differently in the years to come. Once again: that huge boulder has already started to roll.

Over the course of this book I will introduce you to dozens of former pork, lamb, beef, and dairy farmers who no longer want to earn money through the slaughter of animals and who have switched to plant-based farming. These farmers do exist: I have done online research about them and spoken to several of them; I have at times seen their transformed businesses with my own eyes, and everything I have learned in the last few years while researching this book has convinced me that many more will follow their example in the very near future. I will introduce you to monkeys and other animals who have obtained human rights, robot pets, and chefs and restaurant owners who no longer call their menus "vegan" because a plant-based lifestyle has become such a normal thing in their lives that it no longer requires a separate label. I will introduce you to kitchen appliances that measure what nutrients your body needs and therapists who specialize in solving issues between vegan and non-vegan partners and family members. I will introduce you to *vegansexuals* (vegans who only date other vegans), houses that float on higher sea levels, and villages that are hurricane resistant. I haven't dreamt up any of these stories—these things were all taking place in 2019, all across the world, but perhaps you just weren't aware of them yet because they were a little bit farther from home.

Realistic Futurism

It is important that you let this sink in: even though the story I will tell over the following chapters largely takes place in the future, it has not been plucked from the more creative parts of my brain. Everything you will read over the pages to come has already been invented and implemented. It exists, just not on a large scale yet. What I describe is futuristic, but it is also reality.

I am going to show you not only what a more animal- and environmentally friendly world would look like, but also what the consequences would be for our economy, climate, health, and culture once a plant-based lifestyle is more widespread. You will also learn that this new plant-based world is not perfect. In a world where animals are no longer abused for human use, we still have to deal with ethical dilemmas, only these will concern the way we treat living beings other than chickens, cows, or pigs. Or they will concern the shame with which the new generation looks back on our complicity in large-scale problems surrounding animal welfare and the environment. We also won't suddenly all be in perfect health when we switch to eating more plants than animals en masse (you will see that vegan food can be just as unhealthy as eating an antibiotic-ridden piece of cut-price meat). We will also be faced with new problems, such as the loss of certain products and professions, which will require new solutions, and I will show you a number of these.

I will leave it up to you to decide what you make of this future world. I personally must admit that I found it less than ideal (at times I even found it downright terrifying), but in my eyes these new problems do not outweigh the many problems we as humans are facing in the current system. I am unable to reach any other conclusion after conducting extensive research and dozens of interviews with international experts in the fields of livestock farming, foodstuffs, climate, and energy. In a world in which we no longer use animal products—in a world after the Protein Revolution—we will have succeeded in combating the most disastrous scenarios of climate change, and the people of that world will be healthier on average and far, far fewer animals will suffer stress and pain.

The Protein Revolution

This is one of the future scenarios that will quickly unfold if we mobilize ourselves quickly and collectively. Perhaps reading this immediately makes you a little skeptical. There's a good chance your thoughts go something like: "This sounds good, but there's not much sense in me abruptly changing the way I do everything in my life, because the rest of the world isn't doing it and so nothing will happen in the end." You wouldn't be the only one. This is, in fact, the most commonly used argument against any suggestion of changing things in our day-to-day lives. Yet the fact this argument is popular doesn't make it a strong one.

We know from history that social changes that were so vast and so radical that the people who lived through them could hardly even imagine them happened anyway. Even when there were many people against those changes, when few wanted or dared to go along with them initially. When laws on the equality of slaves were first being discussed, skeptics and opponents often argued that people would never accept such radical changes, arguing that such a transformation would seriously threaten the economy and would therefore be a serious danger to society—but luckily this did not hold those pioneers and activists back. The revolution came about anyway, after which the naysayers were suddenly shocked to find themselves among a small group of stuffy, old-fashioned lawbreakers that time had left behind. Once you reach the end of this book, it will be up to you to decide which future you find the most desirable and what role you wish to play in history (for you do play a role: in the game of life there are no holidays or sick days).

The book you are now holding in your hands is a lot more cheerful than most other futuristic books, since they often talk about the end of the world: an end where floodwaters engulf the earth or forest fires wipe out all trees, plants, and other living things. This book, however, is about a new beginning for the world. Your future starts on the next page.

1

How Farmers Can Change the World

On the day that Swedish pig farmer Gustaf Söderfeldt sold all his animals, the grassy meadows around his farm seemed much larger than usual. It was unusually quiet. The stalls were empty. The air bore down heavily onto his shoulders, and he paced, confused, back and forth from his stalls to his pastures. Once he had come back, it was as if he had again forgotten what he was going to do. But he hadn't. There was simply nothing to do. His mind was elsewhere, not ready to think about how his working days would soon be occupied.

It was 2017; he had two young children and little savings. He was worried about the future and only had a vague idea of how he would earn money to provide for his family. Yet even through all of this insecurity and doubt, he felt immensely relieved. "I could have cried. Out of joy, I mean, as I would never kill a pig again. I no longer had to."

It wasn't as if he really had a choice, however. He simply couldn't do it anymore: leading groups of visitors around his organic pig farm and telling them about his "animal-friendly" treatment of pigs. They pointed at the large pasture where the pigs roamed during the day, and in his shop he received compliments about how well treated his animals were, and how this could clearly be tasted in the meat. "I treated my animals better than other farmers did, but that was only relative. In a sense, I lied to my customers, and I lied to myself. I said what I had to say in order to sell my produce, and also to keep feeling good about myself. But I knew what happened when I brought my animals to slaughter, and there was nothing cruelty-free about that."

Gustaf tells me his story from an armchair in the corner of his greenhouse. He shifts back and forth a couple of times in his chair before he continues talking. "The first time that I helped kill a pig at the slaughterhouse, I felt proud. Manly, strong." He is silent for a few seconds, as if debating whether to continue or not. He opens his mouth, then closes it again. Finally, he comes out with it: "It made me feel powerful."

If he looks back on it now, is he ashamed? "Yes and no. Yes, I think it's repulsive what I did to my pigs. It's also terrible to think that I derived some sort of pleasure from it. In general I'm a gentle person—I didn't know I had these kinds of feelings. Perhaps I don't want to know my true self either. But shame isn't the right word. You have to understand that during those early years as a pig farmer, I truly believed what I was doing was morally right. I believed this because I compared my own operation with how intensive livestock farming worked. In my mind, only bad farmers took part in that. It was bad for their animals, bad for people's health with the contaminated meat they produced, and bad for the environment. I was their polar opposite in all aspects. I was the small-scale farmer, with a humane slaughter label on his meat. I was the good guy. So how else was I meant to feel about my decisions?"

Before he became a pig farmer, Gustaf was a city boy. He and his partner, Caroline, decided to move to the country when they were in their late twenties: they wanted peace and quiet, to be in nature more and to work with their hands. Becoming farmers seemed to be the most obvious way of making money out in the country, and all the farmers they knew kept animals. This was good news: both of them were animal lovers, and both of them were horrified by the overpacked factory farm stalls they occasionally saw on television, the crowded trucks filled with animals on the highway, and the antibiotic- and stress hormone–filled meat you found in supermarkets. "We wanted to do things differently. We wanted to keep happy animals, where we could care for them properly, and then kill them in a painless, stress-free way in order to make honestly produced meat from them."

The Pioneers

They sold their house in the city, bought a plot of land in a small village, and used books and courses to learn how to be farmers. They bought a couple of pigs, some sheep, goats, chickens, and ducks, and it seemed like they had found a gap in the Swedish market. People living nearby were drawn to the farm to see these new farmers in action. They came for the animals wandering freely over the land, for the young owners with their idealistic vision, for the old-fashioned small-scale nature of the farm. More and more visitors began coming to the farm and asking for a guided tour, and people wanted to buy their "honestly produced" meat. The meat was expensive, much more expensive than what you could buy at a supermarket, or even at your local butcher. Yet people bought it. They paid extra not just for the taste but also to soothe their consciences. They were eating good meat after all, from good pig farmers. In no time there was so much demand for Gustaf and Caroline's pork that they bought more pigs and opened a shop where they could sell their produce.

Their farming business bloomed. Gustaf, however, began developing a sense of hesitation.

"Over the years, something changed within me. More and more often I [would] look into the eyes of my pigs as I herded them into the truck to take them to the slaughterhouse, and I [would] realize they were terrified. They might have followed my commands when I shouted at them, because they were used to me being the boss and couldn't do any differently. But they resisted in other ways. The way they looked at me. By walking backward when I wanted them to go forward, which meant I had to pull them forcefully over the ramp. By screaming: pigs can make a terrible loud noise when they are scared."

As he led another group of customers around his farm a few hours after a trip to the slaughterhouse, Gustaf often thought about what would happen if he were to talk honestly about the noises his pigs made when they were at the slaughterhouse. "Spine-chilling shrieks," he remembers. "High-pitched, shrill. Mortal terror. They knew as soon as we got there. Of course they knew, they could hear the screams of the other pigs as they waited their turn inside. And they could smell the blood. I smelled it too. You can't escape that smell."

Gustaf looked at his customers and wondered what would happen if he were honest and told them that his pigs would struggle when they got close to the place where they would be slaughtered. How the workers had to pin them down hard as a result. "Or I wondered how my customers would react if I told them I took the piglets away from their mothers shortly after they were born, because that's just how it works in the meat industry, and how the mothers would then try to run after them, and how they panicked when I prevented them from doing that, because that's what mothers do when they can no longer look after their children."

He didn't say anything in the end. "I knew I wouldn't have any more customers left." So he smiled, kept silent, and took compliments that made him more and more uncomfortable. Something wasn't right in his life, but he didn't see how he could do things differently. This is what being a farmer meant, this is what he and Caroline had wanted, this was their livelihood, and in any case they were doing it in an honorable way. Other farmers in the area left the dirty work of slaughter to cheaper, animal-unfriendly slaughterhouses. Gustaf did not. "I didn't want to be like those city folks, who bought prepackaged, sliced, and unrecognizable manufactured meat products. They claim to be against animal abuse, but they don't want to know what happened to the animal that's now on their plate. I also didn't want to be like those other farmers who let other people slaughter their animals. I wanted to take responsibility."

So he brought his animals to the slaughterhouse himself. He shot the pins through their heads himself, or held the pigs down as someone else did it. "The first few times I did that, I got a kick from it. After that, I began to feel worse and worse. More numb . . . but I didn't understand what was happening to me."

The Crisis

Then came that afternoon when, after transporting the pigs, Gustaf walked into the kitchen to find Caroline with a drawn, pale face, a laptop open on the table in front of her. She told him that she had spent hours watching videos about veganism on YouTube. In these videos, activists explained why there was no such thing as humane meat. They suggested that young, healthy animals, like any other beings, don't *want* to die, and so a forced, premature slaughter is always accompanied by a huge amount of stress. Even if they've been well looked after in the years leading up to it. Even if their death happens relatively quickly. Much the same way that a healthy, young human wants to stay alive and would freak out if you tried to kill them, even if you were going to do it as gently as possible. Caroline had clicked from one video to the next, and Gustaf now sat and joined her. They couldn't stop. They spent the whole evening watching them.

"Everything they said fit with what I'd been feeling intuitively for a long time," Gustaf remembered. "I suddenly realized our whole plan to become good pig farmers, to sell humanely slaughtered meat, was all based on a misconception. We had become successful farmers but not cruelty-free ones. Yes, we let our pigs roam freely, and we fed them plenty during their lives. But once they were big enough to make money, we scared them, and killed them, long before they would die in nature, the males after a few months, and the females after a few years, once they had had their piglets. We took their children from them; we hurt them, and ended their lives against their will. And we, 'animal lovers,' we made money from all of this!"

The shock of that realization was enormous. "I suddenly understood why I had begun to feel so numb inside. I wasn't living according to my own values! I thought back to all those times at the slaughterhouse, and I was disgusted by my own actions. If you really wanted to be kind to animals, you wouldn't breed them for meat production. You wouldn't cause them stress by separating mothers from their children. You wouldn't kill them when they could have lived for many more years."

Almost that very same night, the couple decided they didn't want to carry on. Both of them became vegan, sold their pigs, and used the money to develop and carry out a new business plan: from now on they would only grow and sell vegetables.

What Gustaf and Caroline didn't know back then was that all around the world, other farmers had gone through a similar transition. All these farmers, whether in the US, Canada, the UK, Israel, or Germany, had experienced roughly the same psychological process that Gustaf had described in our conversation; from a conviction that they were being a "good" farmer, and therefore a good person, to a nagging feeling that what they were doing was not compatible with their deeper values, until they reached a painful, mostly terrifying conclusion: that for all these years, they had consciously and willingly inflicted suffering on animals, and that their behavior was *immoral*.

That is a bold, and for many people offensive, statement; yet once again this actually reflects what the farmers who have gone through this change felt. Just read what Bob Comis wrote on the blog for his "humane, pasture-raised-and-grass-fed" pig and sheep farm: "This morning, as I look out the window at a pasture quickly growing full of frolicking lambs, I am feeling very much that it might be wrong to eat meat, and that I might indeed be a very bad person for killing animals for a living."

Former dairy farmer Michelle, from Israel, judged herself just as harshly. She had worked for a dairy company since the age of fifteen and married a dairy farmer. There are photos of her in which she—a young, blond, happy smiling girl—is bottle-feeding a calf. In an online interview, she shared that she now cannot look at those photos without getting emotional: "I still live in a state of denial that I used to be a farmer. Everything that has anything to do with dairy farms is very difficult for me. I am not talking about a one-hour visit to the farm . . . whoever is really inside knows what kind of place it is. It's hell. There is terrible suffering there. Cows are beaten, we take the calves away from their mothers, they scream and scream, they resist being milked . . . so we tie their feet down. The screams of the mothers . . . I still hear the sound. It won't go away."

Her trauma seems to be reinforced by the fact that, now that she no longer works on the farm, she can hardly imagine that at one time she would whistle while doing the things that now cause her to weep when discussing them, as if the Michelle from before is a stranger: not just someone she no longer recognizes but someone who, thanks to the power of hindsight, she hates. "When I was a farmer, I burned out horns, which is painful for calves! I clipped nipples (after a calf is born in the dairy industry, any extra teats are cut off from the udders, as they pose an infection risk for when the cow is milked later on), also painful. I sent mothers and their babies to slaughter. I separated babies from their mothers. And somehow I saw nothing wrong with it." This is not so unusual: everything that Michelle did happened fully in accordance with the dairy industry regulations.

Just like Gustaf, she describes the moment she stopped being a livestock farmer as an identity crisis. Developmental psychologist Erik Erikson described an identity crisis as a state that humans go through when they begin to question who they actually are, when the image you had about yourself no longer fits with the person you see in the mirror.

"It was a difficult period," Gustaf says when I ask him about the moment in his life that he describes as an identity crisis. "I was a mess . . . about the company—we had to make so many practical decisions in such a short period of time—but above all about myself. Who had I been all these years?"

But even though he spent that period doubting everything, he still felt better than he did before. "It was as if something was released within me." He points to his chest. "From the night we decided to stop pig farming, I better understood why I had felt so unhappy for all those years while we had been living out our fantasies of rural idyll. I killed my animals because I thought that it was part of being a farmer. Now I knew that this was destroying me inside. It's not right, and it starts to eat away at you, and the only way you can carry on doing it was by burying your feelings. I had become numb, but that is what this job does to you, and now it felt like I had rediscovered my true nature. I wanted to do things better. I didn't care if it would be financially difficult. As far as I was concerned, vegan farming was the *only* way forward."

Michelle had the same kind of experience. "All the sorrow that I caused to them is forever engraved upon my heart. When I think of numbers, I have no idea how many mothers and babies I put on the trailer taking them to slaughter. How many mothers were left without their babies. And they screamed. If someone would touch my daughter or my son . . . I just don't know what to say. Just the thought of it frightens me."

Howard

In Montana, cattle farmer Howard Lyman stood washing his hands in his bathroom. He stared at his face in the mirror. He looked good, so he said himself. A few years prior he had become a vegetarian for health reasons, and started feeling so good that he had recently also scrapped dairy products from his diet. Since then he had felt even better.

Physically at least.

Mentally, things were not so great with him at all.

As a young man, Lyman had been taught by his parents that humans needed meat and dairy in order to stay healthy.

This made him proud of the profession his great-grandfather, grandfather, father, and now he himself had chosen. They were doing important social work by breeding cows, keeping them, and then selling them to the milk and dairy industries. America was hungry and needed to be fed, and his family business was helping to do this. He personally had 7,000 cows. "[Me and] the people I knew in animal production were good people just trying to do the best they knew how for what they envisioned were the right reasons. . . . [We] believed [we] were providing an absolute necessity: first-class protein. It was ingrained in them from the time they were kids: eat your meat. Drink your milk. Stay healthy."

But he had stopped consuming both for some time already. He didn't drink milk or eat any butter or cheese either. And yet he felt better and healthier than ever—a physical experience that didn't sit right with his convictions. Could it be, thought Lyman, that other people also didn't *need* animal products in order to stay alive? Had his parents, schoolteachers, and everyone else who had taught him that animal proteins were vitally important to humans all been wrong? "No way," Lyman mumbled to himself. It couldn't be true, else it would mean that he, his father, his grandfather, and his great-grandfather had been killing animals unnecessarily all this time. That would be intolerable.

But his body didn't lie: it was fitter and stronger than ever. A thought shot through his head: My God, have I been killing my animals for the wrong reasons? "[It] was a door of my soul that I had never opened before," remembered Lyman about that moment in an online article. He shook the water off his hands, looked deep into his eyes in the mirror, and thought of his cowshed, his cows, and the slaughterhouse. The answer that slowly began to take shape in his mind was, in his own words, "so traumatic for me that I damn near tore the sink off the wall."

The door into his soul did not close, despite all his efforts to do so. How was he meant to bring the family tradition to an end after four generations? How was he meant to open the bathroom door, go to his wife, and tell her that he had realized that he had to shut down their million-dollar business as soon as possible? What would he say to her? "Wait a minute: I think what we are doing is wrong"?

Howard stared at the lines around his eyes, the wrinkles on his forehead. He was scared. He realized their livelihood was built on shaky foundations and could collapse at any moment. "Everything I'd believed in my entire life was at risk because there I was with a business built on killing animals."

But just like Gustaf and Michelle, the panic Lyman was now dealing with was accompanied by a profound moment of clarity. The only thing he could do was stop.

"I knew what those animals looked like when they went onto the kill floor," he says. "I knew what was in their eyes, and I was the person putting them there. Try to imagine the appalling situation a farm animal finds themselves in when they are about to be slaughtered. A hugely sensitive and intelligent being, a being that knows what's going to happen, trapped and unable to escape: just try to really imagine what that must be like for an animal."

He opened the bathroom door and called his wife.

The Ripple Effect

If you watched a slow-motion film of a drop falling into some water, you would see that immediately after the drop has touched the surface, a very small well is created, with a ring of small droplets around its edge. This shape is just about visible to the naked eye, and you'd miss it if you didn't watch the film in extremely slow motion. Within six-hundredths of a second, the drop jumps back up out of the well, and then pulls a thin trail of water out behind it. Now, imagine you're observing this in slow motion. You would see how the original drop broke free from that trail, how it very briefly sits, reborn, above the liquid, before finally falling back down to become one with the body of water, and soon the water's surface is flat and undisturbed once more.

The identity crisis that these farmers experienced somewhat resembles what happens to a pool of water when a drop falls into it. The realization (the drop) that they might have spent years doing something that goes against their core values creates a hole in the image they have of themselves and their chosen profession. That realization comes so unexpectedly that people experience it as trauma. It drags them down, into the depths; a dark place, filled with shame, guilt, anger, and misunderstanding. Then there comes a moment when the droplet jumps back up again, toward the light. At this stage, relief triumphs over making a decision, the radical break with a past that, upon closer inspection, does not fit with who they really are. The farmer sees the opportunity to do things better, to live in accordance with their deep-held values, and if that succeeds, the surface of the water will be calm once again: the image someone has of themselves once again fits with what they see in the mirror.

Nowadays Gustaf grows legumes and vegetables, and also gives agricultural studies lectures, where he tells future farmers that having to keep animals in order to survive as a farmer is a myth. "We were also uncertain when we switched from keeping pigs to vegetables," he says to packed lecture halls, "but it turned out we could make a good living from it. Our land is also fertile without using animal manure, and we're now fully convinced that the future of farming is growing vegetables." He is enthusiastically bringing other people into the fold. "We often get visits to the farm from young people who want to get away from the city, just like Caroline and I once did. They all want to grow fruit and vegetables, just like we do now. This is the kindest way to live, and it's smart too: you save huge amounts in terms of costs when you don't have to keep animals."

Former cattle farmers Michelle and Howard Lyman also share their story with other farmers. In interviews, books, lectures, articles, and videos they tell whoever wants to listen about the realization they had. Lyman even made it on to *The Oprah Winfrey Show*, and after hearing his story, Oprah declared she would never ever eat a hamburger again. Former sheep farmer Bob Comis gives workshops to vegan farmers on his now-transformed farm. Former goat farmer Susana also does this: she enthusiastically tells visitors how she now no longer makes cheese from goat's milk, only from hazelnuts, and she feels much better as a result. "Going without dairy or meat is not hard for me anymore. What's hard is seeing people stuck in a mindset where their daily decisions are fundamentally, violently at odds with their most basic values. We can and should do better."

Psychologists would call what these farmers are doing "corrective actions." This is something positive that someone can do which is somehow connected with trauma. Carrying out these "corrective actions" results in a positive self-image. By helping and educating others, these farmers derive meaning from the shame they feel over what they now perceive as their past wrongdoings, and at the same time their activism is creating a ripple effect. Drops become rings, and rings become waves.

Jay and Katja

They are still taken aback by how quickly everything went. One minute they hesitantly said "yes" to a question from an already converted farmer about whether they wanted to stop farming animals; the next they had got rid of their cows and stopped being cattle farmers. "I really miss them a lot," Jay says solemnly. He is middle-aged, balding, and wearing glasses and a green sweater. "But now they live in a cow sanctuary, where they're having a great time, and that makes me feel good." We are meeting via Skype, with Jay in front of his webcam, his partner, Katja, just behind him and to his left. Whenever he forgets something or can't find the words to express a thought, he glances over his shoulder at Katja, who proceeds to finish his sentence in a loud voice. Of the seventy cows that the couple owned, twenty are still with them. "They'll stay with us until they die," says Jay. "They give our land manure so we can grow crops here, and in exchange we give them good food and loving care. I will never take their milk or meat from them ever again. Our relationship with our cows is now fairer than it used to be."

The "used to be" Jay is talking about had finally unraveled just a few weeks before this conversation. The process began after they had been visited by a fellow British farmer who had been making a living from growing vegetables for years, and who tried to convince them to give up producing animal products and start growing vegetables. That was the future, he had said, and let's face it, wouldn't it be nicer if they never had to harm their animals again?

That question really hit home with Jay and Katja. "For a long time, we felt the way we had treated our cows did not fit with what we really wanted for them. What that farmer said seemed so . . . true. In all honesty it was very difficult, because it meant that up until that point, we had been doing things wrong." Their story resembled that of the other farmers I spoke with for this book, except theirs is so fresh and new that they still often struggle to describe the enormous changes they have gone through since the visit of that vegan farmer, both to their daily lives and to their emotions. You might say that in their case, the drop is still moving as we speak, from the depths of the water to the space just above the surface. Jay and Katja largely feel relieved, but also sad: there is concern, but, at the same time, a cautious hope for the future.

"I loved my cows," Jay says hesitantly.

"Enormously," adds Katja.

He nods. "I thought I was treating them well. But there was always that feeling of guilt when they had to go to be slaughtered, or when we had to take the calves from them. Or rather, it was a suppressed feeling of guilt. I now think I didn't understand it because I saw no other way out. Am I explaining that right, Katja?"

Now it is her turn to nod. "It was too painful."

"Yes. So I told myself that there was no alternative, and it was normal to treat animals this way. That this was my job, as a farmer." But when that activist farmer who visited them claimed they didn't need to keep animals if they wanted to be farmers, let alone killing them or causing them stress, the feelings of guilt that Jay had been repressing for years all suddenly came to a head. "I knew right away that I had to get rid of my cows, I knew they deserved to get away from me." The vegetable farmer promised to help the couple find a sanctuary for their cows, but he warned them it could take months to find a suitable place. Jay and Katja were glad for the delay, though: this would give them enough time to come up with a new business plan, apply for conversion subsidies, and mentally process the fact they would be saying goodbye to their years of existence as dairy farmers.

Barely a week later the phone rang. Dozens of vegans had voluntarily helped in the search for a sanctuary, the farmer told Jay enthusiastically, and a place had been found where the cows could all grow old. When could they come pick them up?

"It's better this way," says Katja, more to her husband than to me.

He agrees. "I betrayed my cows. They trusted me. For years they let me take their milk from them, and then I hurt them. What we did to those friendly animals is unspeakable. I don't understand why I didn't realize this for so long, but when I look back on it now . . ."

He looks upset, shrugging his shoulders slightly when I ask him to finish his sentence. Only when I ask him about the twenty remaining cows on his farm do his eyes light up again.

"My bond with them now is so different, it doesn't compare," he says. "It's as if now I can really feel love for them. Or does that sound kind of new-agey, Katja?"

Resentfully she answers, "No, that's exactly what it's like! Now we don't have to push down these feelings anymore."

Jay again: "I am no longer constantly obsessed with their impending demise. That's all I would think about before. Somewhere in the back of your mind, you know you shouldn't attach yourself to an animal, you shouldn't identify with a cow, because in a few years you'll have to kill it. So you tell yourself a cow won't be affected if you remove her child from her, or a young bull barely feels a thing when he's being slaughtered. You just do it and try not to think too much about it."

Katja, with a sigh of relief, says, "But thankfully that doesn't have to happen anymore."

#mindfuck

In this selection of stories from farmers that I've shared with you, as well as in the many stories of other farmers that I haven't shared (because I still have so many more things to tell you), the same patterns always seem to emerge. The farmers identified themselves as animal lovers throughout their whole lives, and they treated their farm animals in a way that they would have described as "loving" and "good," until something in their lives happened that radically changed their view of themselves. They suddenly saw themselves as "animal abusers" and, worse still, as people who did something for money that, according to their own value system, was immoral. How was this possible? How can it be that one moment you are fully convinced that you are doing something good, be it for the world, the animals, or your family, and the next, within a week, day, or hour, you have decided that your entire lifestyle is evil? How can you go from spending decades being proud of your work and the way you do it to suddenly being so ashamed of it that you can no longer talk about it without bursting into tears? And how can it be that, in the twenty-first century, people from all over the world are realizing at the same time that they have been doing things wrong this whole time?

This is the result of a phenomenon that I call the "mindfuck of our age," and which I will explain during the course of this book. A mindfuck is a phenomenon that spiritually confuses you or misleads you. This is what happened in the twenty-first century to these farmers, to me, to you, and to millions of other people.

2

Why Good People Believe in Bad Stories

I'm willing to bet you know the story of the Big Bang, one of the most popular stories in the Western world. We pass it down from generation to generation. As children we are taught it at school and by our parents, and later we tell it to our own children. The words we use are different from person to person, but in general this is the story we all tell: 13.7 billion years ago, something came into existence from nothing. First there was no Earth and no sky, no time, no darkness, and no light, but then suddenly the universe got bigger and bigger; stars were formed, then planets, including Earth, and on Earth life appeared. The moment of expansion in this story is widely called the Big Bang, and the story itself the Big Bang Theory.

Well, I don't know about you, but when I read this, I think the story is a load of bull . . . well, manure. The story has a somewhat misleading name, as scientists are not entirely sure if there actually was a bang, and if there was, it certainly didn't take place at the beginning of the story. The universe is constantly expanding, which means we are still in the middle of this explosion. "The ongoing primordial rumble" would therefore have been a better name. Yet this more accurate, less melodious name provokes more questions about the story than it provides answers. It feels only half-baked: the second part of the story follows some kind of logic, but that botched, implausible beginning is something even a novelist wouldn't get away with. Where did that nothing before the beginning actually come from? And why was it there? Where is it now? And what should we imagine? Nothing?

If at this point you feel like complaining about this shaky story to the scientists who came up with it, they would probably agree with you. They would smile at you self-deprecatingly, shrug their shoulders, let out a sigh, and then admit to you they also find it a relatively flawed story. They would add that while it might make sense in broad terms, it is something so very complex that it goes beyond human comprehension, and that is why we never quite succeed at telling it. Yet we stick to it, because we haven't yet come up with a better story.

People like a story. A half-finished, somewhat illogical story is better than no story at all. Stories about who we are, where we came from, and why we do what we do give people a sense of security. If we can explain what happened in the past, it makes predicting what will happen in the future seem easier. We are then at least not entirely at the mercy of chance, or fate, or whatever name you want to give to everything we do not understand.

And so, this is what we currently do with our feeble story about the Big Bang until someone comes along with a better story and we rewrite the history of the universe. Only then, once we are content with a new story, will we all be able to look back at that old-fashioned Big Bang Theory and say, "It was pretty poor, wasn't it?"

Our Idiotic Ancestors

Something similar happened with another story that has been hugely popular for a long time: the story of the origin and evolution of humankind. This had to be changed in the history books several times, and afterward, these earlier versions were much derided. For a long time people told each other they were descended from a hominid that lived in East Africa around 200,000 years ago. There was sound scientific evidence for this: fossilized human remains dating from that time were found in Ethiopia and were compared to DNA from people from all over the world. Match found, happy scientists, front page headlines.

For years this was a great story to tell down at the pub. That was until in the spring of 2017, when a team of paleontologists discovered that modern humans had existed 100,000 years earlier than had been thought, and in a different place entirely.

Oops.

They found the remains of *Homo sapiens* spread across most of Africa, including present-day Morocco. It must have been a confusing month for scientists, as well as bar-propped raconteurs, because a few weeks prior to that, another spectacular discovery had been made that also undermined the evolution story: people once thought that *Homo sapiens* was our only humanoid ancestor, but now it seems that we had all different kinds of great-great-great-great-grandparents all over the place.

This was major news, as our popular idea about human evolution was suddenly rendered implausible. You will be familiar with the image, one mostly shown as a diagram of a row of human-like figures, the first a chimpanzee, the last a human. The figures gradually become more upright and their heads get larger and larger. The idea behind the diagram is that there is a functionalism in our family tree that could only have ended in our supremacy: we started as animals and then steadily became bigger, smarter, and more human. Yet new discoveries make it clear that this view doesn't add up.

Meanwhile, scientists think that our evolutionary story was a lot more chaotic: 300,000 years ago, different groups of *sapiens* lived on the African continent, including *Homo naledi*. At the same time, Neanderthals were roaming Europe, Denisovans were in Southeast Asia, and the dwarf-like *Homo floriensis* was to be found on the island of Flores. A number of these hominids were more ape-like and had smaller skulls than *Homo sapiens*, but Neanderthals actually had larger brains. We modern humans are descended not only from animals but also from other hominids, because there is evidence that *sapiens* had romantic trysts with Neanderthals and Denisovans, and maybe even *Homo naledi*.

And writers of academic books are already adapting their texts accordingly. For example, a 1950s Dutch encyclopedia contained an entry stating that a Neanderthal's skull had a pathological form "akin to that of an idiot." This text was later fine-tuned, as DNA research made it clear that we carry around genetic material from these "idiots" within us.

Human Experts

I'm not telling you all these stories in order to throw shade at scientists; they were just doing their work with the information they had available at the time. The purpose of these stories about the Big Bang, human evolution, and our "idiotic" forefathers and -mothers is to show you that our ideas about what is "true" are constantly changing—and rightly so, when the science is supported with "facts" and "evidence." We trust the knowledge of experts, and conveniently forget they are people too, with research methods that are too limited to understand things we want to understand in their entirety. Past experience has also shown that we are only able to spot what is wrong with a story when we have obtained new information and have moved on to an alternative one. As long as we have no alternative, we are not actively looking for a new story. What we have read in our school textbooks seemed logical, true, and reliable. Only in retrospect does it become clear we only believed this because we didn't know there was something better waiting around the corner.

This same principle applies when we analyze stories about recent human behavior, except this analysis can suddenly get a little uncomfortable. Shifts in our beliefs about what the "truth" is are only usually funny when it comes to stories that are very distant from us, such as the beginning of the universe and the evolution of humans. As you read those not so well-defined stories from history, you could think, "Well, we were only a few decades or a few thousand yards out—what difference does it make?" We've learned from our mistakes and now we know better. This is different, however, when we look at stories that played out not so long ago, with lead characters that resemble ourselves, who believed in truths that we know were dubious at best.

The Dark Years Before the Enlightenment

Let's take the Enlightenment. Around 1700, beginning in Britain and France and spreading from there, three important new ways of thinking emerged that are still taught in schools today. These would later go on to determine how our society was shaped: views on tolerance, common sense based on logic, and the need for the equal treatment of people.

All this now sounds obvious, but at the time it was revolutionary. This meant you could no longer burn witches: Our newfound wisdom taught us that witchcraft was impossible, so why would you burn innocent women at the stake? It took many years (and many more burnings) before this new view had spread all over Europe: the last "witch" in Germany was burned in 1749, in Switzerland in 1783.

Other things people had to get used to included the idea that beggars should no longer be beaten in the streets, non-believers should no longer be expelled or imprisoned, young girls should no longer be married off to old men they didn't know, and peasants were not brought into the world for the sole purpose of working for a wealthy landowner and otherwise keeping their mouths shut. These were all ideas that at the time were very common among people from all walks of life and all different countries.

As you read this, it is hard to imagine that people back then really believed what they were doing was right. Yet this is what they thought, and in any case, the vast majority of people were unaware they lived in an unenlightened society. They burned witches and nonbelievers because they wanted to protect their families from "evil," they beat vagrants because they found them dirty and dangerous, and they tried to cure illnesses with superstition because they had been taught that this would help a sick loved one. Only in retrospect, when it became the norm not to do all these things, when laws were made that banned people from doing what they had been able to do all this time, did people begin to look at their old behavior with new eyes and create a new story about what was just and reasonable. In history books, this alternative story would later be called "the Enlightenment."

In his book *A Little History of the World*, Ernst Gombrich gave a lovely description of how over time we can think so differently about what is "good": "Have you ever come across an old school exercise book . . . and, on leafing through it, been amazed at how much you have changed in such a short time? Amazed by your mistakes, but also by the good things you had written? Yet at the time you hadn't noticed that you were changing. Well, the history of the world is just the same. How nice it would be if, suddenly, heralds were to ride through the streets crying: 'Attention please! A new age is beginning!' But things aren't like that: people change their opinions without even noticing. And then all of a sudden they become aware of it, as you do when you look at your old school books. Then they announce with pride. 'We are the new age.' And they often add, 'People used to be so stupid!'"

Something similar happened during the era of the Enlightenment, and later during the era of slavery and women's rights, and later still during the shift from carnism to veganism.

The Sleeping Cat and the Roast Chicken

If, like me, you were born in 1983 in a Dutch city called Utrecht and grew up in a loving family with a dog who answered to the name Kaj, a small gray cat called Klauwtje ("Clawy," due to her habit of scratching so much), and a much fatter ginger cat called Woutje (because it rhymed with the first cat's name), you probably called yourself a real animal lover. You wanted to become a vet when you grew up; you wanted to make all sick animals better again, and ideally you would have made the cats immortal, because living without them would be terrible. You earned some pocket money by walking a neighbor's dog. In the mornings, before taking the dog out for a walk, you ate brown bread with cheese and drank a glass of milk; after that you put jam or chocolate sprinkles on a slice of bread and washed it down with a glass of orange juice, which you personally found tastier but you were told it was less healthy than dairy. In the afternoon you ate sandwiches with butter and ham (delicious) or salami (even more delicious), and in the evenings you ate spaghetti with meatballs in tomato sauce, or roast chicken, your favorite meal, especially with a crispy skin, dripping with juices. At night you lay in bed with the cat next to you under the sheets while you secretly read a book with a flashlight, and your parents pretended not to know you weren't asleep yet. When you finally couldn't keep your eyes open any longer, you carefully put the flashlight and book away, so as not to wake up the cat.

If someone had told you when you were a kid that dogs and cats were scared, kicked and beaten, fattened up so much they could hardly walk, forced to spend their lives in small cages they could barely move in, and killed when just a few months old and in full health just so you could eat them, you would have broken down in tears and refused to touch any animal meat your parents served you. The idea would have made you angry and nauseated: Who would want to eat creatures as kind as these animals?

Yet this is exactly what happens to the cows, pigs, and chickens that you tuck into on a daily basis, and you didn't think there was anything wrong with that at all. There is, therefore, a distinction in your life between animals kept as pets (which you have to treat with kindness and care) and farm animals (which you can let get sick, kill, and eat). This discrepancy is widespread among people of my generation, and can only be explained if I tell you another immensely popular but logically shaky story: the story of carnism.

Carnism

You, I, and many generations before us had clung to an ideology that would later be called carnism. In carnism, children are conditioned from a young age to eat meat and drink the milk of animals.

The animals you can eat and the ones you can't are determined by the types of animals farmed in the country where you grew up, as well as the cultural myths connected to them. In most Western countries it was completely normal to eat (and mistreat) chickens, pigs, cows, goats, horses, and fish, but eating cats, dogs, and hamsters was taboo. This latter group of animals is seen as people's "housemates" or "friends"; the former are considered food and objects to be exploited. In other countries these categories were different: for example, in China, South Korea, the Philippines, Thailand, Laos, Vietnam, Cambodia, Nigeria, parts of Indonesia, and the Nagaland region of India, eating dogs was very normal, while eating cows in India was taboo because they were considered sacred.

People living in the carnist age were unaware they adhered to an ideology, and in general they didn't realize they were doing anything wrong. Just as people before the Enlightenment didn't realize they had been living in the dark for decades, people back then did not realize they had been living under carnism.

That was because the story of carnism was the dominant ideology of that age. When an ideology is dominant—that is, when it is a system of beliefs held by the vast majority of people—it is very difficult for individuals to recognize this as a personal and voluntary *choice*, as well as it being an idea that you can choose to believe in . . . or not. Yet carnism certainly is an ideology. If eating meat is not a necessity for survival (which is the case for most places in the world), then doing so is indeed a choice, and choices are always the result of a belief: you don't have to, but you do it anyway because you think you have a good reason.

The Three Myths of Carnism

We animal-eaters believed we had three good reasons to live according to the ideology of carnism. Psychologist Melanie Joy, a Harvard graduate, says that a number of popular myths circulated that legitimized carnist ideology, and therefore that eating meat was *normal, natural*, and *necessary*. These three stories were actively promoted by doctors, scientists, dieticians, teachers, and parents: they were told so often, and by so many people, that they seemed irrefutable.

It was a long time before these myths were debunked. First, nutrition and health experts discovered that eating meat was not as healthy as had been thought for so long. Certain kinds of meat were increasingly linked to diseases that at the time were as common as they were fatal, such as cancer and obesity. Scientists additionally discovered that people who ate meat-free diets were generally healthier and fitter than people who ate lots of meat. If meat is not healthy for people, this also suggests meat is not necessary for our health.

Then came the news that primitive man wasn't a meat-eater. Eating meat is therefore not per se "natural." *Homo naledi*, which I briefly mentioned earlier, appears to have been vegetarian; early *Homo sapiens* didn't go around killing great big animals with clubs, but rather scavenged the meat off the remains of the prey killed by large predators. In fact, the great majority of modern humans ate mostly grains and roots well into the Industrial Revolution: for a long time, eating meat was something only done by rich men. Poor people couldn't afford it, and furthermore it was claimed to be unsuitable for the female constitution. Thus meat has not been a "normal" foodstuff for the majority of human history.

I will expand on these discoveries later, but for now the most important thing for you to realize is that people who believe in carnist ideology (animal-eaters) are not violently predisposed, but simply behave this way. If eating meat is not normal, natural, and necessary, then the choice to do so is an extremely violent one. You cannot produce meat without violence, and egg and dairy production harms animals. Most animal-eaters, however, would say they have values such as compassion, empathy, and justice, and this is indeed true when it comes to other people and certain animals. The only instances in which these values do not apply are for what animal-eaters consider edible animals.

A large-scale study from 2017 showed that over 70 percent of Americans believed that people have a duty to look after animals properly. That same conviction was shared by the farmers in chapter 1, who described themselves as animal lovers (their love for animals often being the very reason why they wanted to become farmers in the first place!), but who also found it very normal to stress out, hurt, and kill the animals on their farms. This conviction is also reflected in my memories of my own childhood and the way I cared for pets, while at the same time eating the flesh of farm animals.

In short, carnism can be defined as an extremely violent

ideology that millions of peace-loving people passionately believe in. They continue to cling to their ideology while it goes directly against their core values. How can intelligent, mild-mannered people behave in ways that go against their deeply held beliefs, yet still feel good about it? The answer to this question consists of a number of different factors. I already mentioned the first of these using the examples of the stories of the Big Bang and evolution: people are inclined to hold on to stories with a shaky foundation and logic for lack of an alternative. The violent lifestyle of carnism is a given that people get used to from birth: this process of familiarization makes living in a radically different way seem impossible, and so the carnist way of life becomes the right way of living. I will explore the other elements of this answer in the last part of this chapter, specifically the mass deception of objective reality by those in power from the carnist age, and the constant denial of objective reality by our individual minds.

Deception

Reality check: today, meat, dairy, and fish companies are often portrayed as small-scale, family-run businesses, but in fact they have been part of billion-dollar industries for a long time now. While the number of vegetarians and vegans has increased enormously since 2018, the annual figures for meat producers have also grown during the same period, particularly because the market for meat in non-Western countries such as China and India has increased. Between 1961 and 2018, global meat production almost quadrupled from 78 million tons to

340 million tons per year. In the summer of 2018, the website Beef2Live (with its slogan "Eat beef, live better") declared that global meat production had had a record year.

Huge amounts of money have also been made in the dairy industry. Even as demand for nondairy alternatives in the Western world has been rapidly increasing, causing a drop in revenue generated from traditional dairy products, dairy imports to Asia have increased, offsetting those lost revenues at home. Cheese is doing particularly well, as it is considered a luxury product, and increasing affluence, particularly in China, is boosting demand for this type of gourmet commodity.

Then we come to the fishing industry, which is booming like never before, and a brief description of this will demonstrate how unfriendly to animals and the environment our whole food system in the twenty-first century has become. This notion isn't shared by everyone in my social circles: when they hear the word "fishing," many people still think of the relaxing activity that their dad, grandpa, or elderly neighbor used to occasionally talk about. Head down to the stream at the edge of town, cast your line into the water, hours spent bent over a motionless patch of water until suddenly, finally, a fish bites, just big enough to stick on the grill, just big enough to be lunch for a family of four.

It hasn't been this way for a long time.

The fishing industry has become a vast, hyper-technological mega-industry that uses scanners, huge nets, and automated catching techniques. One common method, for example, is the gillnet, a net that drags behind the fishing vessel for hours or even days, causing constant damage to the fish that get caught and dragged along with it, while they are still alive. The other commonly used technique is fishing with dragnets, catching entire schools of fish and squashing them together in the small confines of the net until they are finally dragged on board, to slowly suffocate among the mass of their fellow fish. Furthermore, dragnets damage the coral formations encountered in their way, and often snare—and damage or even kill—other marine life caught along with the fish.

What's more, many of the fish in these nets are caught for no reason—they are part of the by-catch, which is usually vast. In 2010, for every one portion of shrimp caught in Dutch waters, one to five portions of other fish species were thrown back dead into the sea. For tropical shrimp, such as those often caught off the coast of Suriname, this by-catch can be up to 90 percent. For another Dutch favorite, sole, the by-catch can be upward of 70 percent. This means that for every 100 pounds of fish caught, 30 pounds of sole are landed and sold, while 70 pounds of other fish are thrown back into the sea dead. Fish stocks across the world are plummeting as a result, as the oceans are literally being cleared of fish. Around half of all sea life eaten around the world does not come from the ocean but from fish farms, where producers try to provide as much fish as cheaply as possible. They do this by keeping fish and shrimp crammed together in nets, cages, or trays, using medicines, hormones, and pesticides. The results of this are devastating for both the health of the fish and biodiversity and the environment. Fish farms produce almost as much harmful emissions as cattle farms, in some cases even more. The water containing so many fish at the same time gets polluted very quickly, and so fish farms are often teeming with parasites, and despite the water being pumped with antibiotics to prevent this, 20 to 40 percent of farmed fish die of illness and infections.

Advertising

Numerous scientists believe that meat, dairy, and fish producers have more influence on people than politics does. For example, they have enormous public relations budgets that they can use to sustain the carnist ideology. They do this by producing attractive commercials that are specifically aimed at encouraging children to eat meat or promoting cow's milk as an essential part of our diet.

In the 1990s, as I was spending most of my energy petting dog and cats, and my energy levels were being boosted by eating farmed meat and dairy, I must have heard the well-known slogan of the Dutch dairy industry on the television on a relatively frequent basis: "Milk: The White Fuel."

Another way that the meat and dairy sectors try to encourage carnism is by ensuring both young and old meat consumers know very little about how meat is actually produced. Slaughtered chickens are cut up into blocks and sold as unrecognizable "nuggets," while pigs and cows are ground up and pounded down, the "hamburger" being the most popular version found in supermarkets, and which no longer look anything like the animal that died to make them.

During the period I was doing research for this book, 95 percent of meat sold in supermarkets and by butchers came from factory-farmed animals, but the meat and dairy ads I saw during that period gave a rather romantic image of preindustrial farm life. For example, meat packaging will often depict green meadows with cows and sheep grazing, even if the animals the meat has come from have rarely ever been outside during their short lives. On packaging for milk and yogurts you don't see images of huge, infected udders and rotting hooves—two conditions frequently found among Holstein-Friesians, the most common dairy cow in Europe and the United States, which was specially bred to produce unnaturally large quantities of milk.

Meat, fish, and dairy produce use misleading labels on their packaging to reassure customers that the rumors they hear about large-scale (and environmentally unfriendly) animal industries are false, and the message is that they are still, at their heart, just a family business, where each animal is known by name, treated with respect, and allowed to die peacefully.

Limited Free Range

The labels used for chicken are a good way of seeing how animal-eaters are misled. The notion of a "free-range chicken" and their "free-range" eggs lets consumers think the animal has had a better life than so-called battery hens or broilers. While this is true, this definition of "better" is rather relative, except this is nowhere to be found on the packaging. Broilers—which you eat in fast-food joints and certain restaurants—are specially bred to grow and fatten in a very short period of time. This often means they cannot walk very well, but they are ready for slaughter within six weeks, when they weigh 5.5 pounds. Up until then they live in a large barn with nineteen chickens per ten square feet, less room than a cauliflower gets growing in a field. Because of the high likelihood that these packed-in chickens will hurt one another or get sick (both of which mean losses for the farmer, as meat from chickens that are sick or die prematurely cannot be sold), they are given preventative antibiotics and the tips of their beaks get burned off.

Without anesthetic.

One thing to note: a chicken's beak is its most important sensory organ.

Poultry farmers did not remove the cause of suffering (too many animals in too little space), but chose instead to adapt the animals to it. This also isn't mentioned on the packaging. Not even in the small print.

So the label "free range" gives the idea that this chicken had a slightly better life than that of its broiler cousins. People who buy "free-range eggs" and "free-range meat" often think that a "free-range chicken" has been allowed to wander around outside at her leisure, scratching and scrabbling in a large space. In reality, a free-range chicken shares a 43-square-foot space with nine others. Unlike broilers and battery hens, free-range chickens can indeed go outside, but with hundreds of others in a huge, closed-off barn with "chicken scratch" spread on the floor. Free-range chickens sometimes have an opening that leads outside, but this is not a requirement to get the certification. Just like broilers, free-range chickens are also given preventative antibiotics, their beaks are burned off, and they grow so quickly that slaughtering the birds after around nine weeks has become lucrative for poultry farmers.

Banned

Similarly misleading labels exist for other highly intensive and animal-unfriendly farming practices. In the United States, for example, animal welfare certification agencies offer three different "humane" methods of castration for piglets and steers: surgery, a technique using rubber rings, and another in which the testicles are crushed—all without anesthetic. Furthermore, under the so-called 28-hour law, introduced by the US government in collaboration with the livestock industry to provide farm animals with "humane protection," vehicles transporting animals to slaughter are required to stop and let them eat, drink, and move every 28 hours. This means animals can be crammed together in a moving truck, exposed to the cold and heat without food or water, and this is considered "humane" *as long as it does not last longer than 28 consecutive hours.*

Most animal-eaters allow themselves to be easily misled by these labels because they simply never come into contact with the large business operations where animal products are manufactured. They do visit smaller farms, where animals are kept as an attraction or for educational or commercial purposes, such as petting zoos and goat farms, which are found in many towns and cities, but these are nothing like the average large-scale chicken, pork, or beef production companies. These are strategically built in sparsely populated areas, out of sight and earshot of animal-eaters.

Even if a skeptical animal-eater wanted to visit one of these mega barns or a slaughterhouse, this isn't such an easy thing to do: these are sealed off from outsiders, and climbing over the fence (to do some filming, for example, which animal rights activists still try to do) is illegal. In the United States, animal rights activists can be declared terrorists if they violate the Animal Enterprise Terrorism Act. In Europe there are other, no less effective methods, which led to a series of public service announcements on Dutch television warning the public about "animal extremism."

Effective marketing techniques by the meat and dairy industries, however, are not the only reason animal-eaters remain true to their carnist ideology. One fanatical supporter of carnism can be found within our own body, without animal-eaters even being aware of it: our brain.

Brain in Control

Our brain is a kind of personal bodyguard. It does its very best to protect us from complicated, painful, or confusing information. It does this so we can keep feeling happy and calm. Psychologists have discovered that we feel our best when we act in a way that corresponds with our beliefs, and when our beliefs, ideas, and opinions come together to form a coherent narrative. Beliefs, ideas, and opinions that conflict with our own, or when we act contrary to our own beliefs, lead to our experiencing an unpleasant tension. In professional literature, this tension is known as "cognitive dissonance." Luckily, however, we are all equipped with a sturdy bodyguard that kicks in as soon as tension increases in our body. Our brain has trained itself to take various emergency measures in order to reduce this dissonance.

One coping strategy frequently used by the brain involves constantly adapting our beliefs to new information. It does this so quickly that we aren't aware of it, with the protective walls it builds hardly letting in any new information that could threaten or upset our internal calm.

Studies have shown that animal-eaters systematically underestimate the intelligence of animals they have been taught from a young age to categorize as "edible," compared to the intelligence of animals they are not used to eating. This opinion continues to exist among animal-eaters, while twenty-first-century scientists have shown that pigs are at least as intelligent as primates (for example, they can learn to use a computer joystick); they experience emotions such as fear and happiness; they like to play like dogs; they are social; they can grieve for fellow pigs who have died; they prefer eating with other pigs than eating alone; and, just like humans, they are able to empathize.

The evidence is all there, and it gets written about by journalists in widely circulated news sources and talked about by television presenters, and yet the implications of this information still don't seem to get through to most animal-eaters. Their overzealous bodyguard stops this information in its tracks in order to come up with reasons why it isn't such a bad thing for a pig to live in a mega barn, to never be allowed to play, to be taken from their mothers too early, or to be slaughtered at an extremely young age. Then we say well, okay, they may be smarter than we thought they were, but they have no perception of time: or rather, "pigs don't understand what 'death' is, so making them experience it is not so bad," or "we can't do it any other way, because people need meat to stay healthy." All of these reasons are repeatedly undermined by new discoveries by animal scientists and health experts, but this does not outweigh the human need to act according to our beliefs and to be free of cognitive dissonance. The only way to do this is by changing behavior, i.e., stop eating meat, but this takes a lot of effort, and getting used to, and discomfort, especially in a society where lots of people around you continue to eat meat. The other way to prevent dissonance is to adjust your beliefs to your behavior. Animal-eaters see themselves as good, friendly, civilized people, and good, friendly, civilized people wouldn't let intelligent, sensitive beings suffer unnecessarily. In their minds, therefore, pigs have to stay stupid and incapable of emotion, and meat has to stay necessary for human health.

Score one for the bodyguard.

Exceptions Throughout Time

Our brain also has another intelligent strategy it employs time and time again to counter that annoying tension: labeling difficult information as an "exception." Our intelligent brain thinks animals have a right to be treated well, except for certain animals that have been put on this earth specifically for our use. Our mental bodyguard has applied this trick not only to carnism but to other ideologies and periods throughout history.

Until two decades into the twentieth century, it was widely believed that women should not have the right to vote because they weren't rational enough. They were "different," and thus were an exception.

In 1865, many white people believed Black people were naturally less intelligent, less industrious, less reliable, and more violent than whites, and consequently could not benefit from an education or hold a good job. You would think that people would have soon discovered that these ideas were unfounded, as Black people could easily prove they were in fact perfectly competent, clean, and law-abiding, but then you would be underestimating the bodyguards we have in our head, and with that the staying power of this kind of dominant ideology: they appear to be remarkably resistant to alternative arguments and ideas because believers end up in a vicious cycle of subjective belief and objective proof. Considering that most good jobs were indeed held by white people, the idea that Black people were inferior would continuously be reinforced. The average white person would have said something along the lines of "look, Black people are no longer slaves, yet there are still hardly any Black professors or judges. Isn't this clear evidence that Black people are just stupider and less hardworking than us?" Black people were therefore not hired to the best positions, and the "proof" of their inferiority was the lack of Black people in high-level positions of power.

This "exception tactic" remained effective for a painfully long time, until 1958 when a Black man by the name of Clennon King applied to study at the University of Mississippi and was committed to a mental asylum. The judge made his decision because he reasoned a Black man who thought he could try to do such a thing must clearly be mad. White superiority was, in his eyes, normal, necessary, and natural.

Sound familiar?

The reason behind this detour in my story about carnism is not that I want to compare the horrifying, degrading way in which Black people have been treated by my white ancestors with the experiences of animals during the age of carnism. My point is to show how stories that are not logical, and that sometimes run completely counter to our core values, can continue to exist for years, decades, and sometimes even centuries. They are carried forward by powerful figures in society, by cultural ideas, and by internal ideas we are largely not aware of ourselves.

Abnormal

At the time of writing this book, there are a large number of people who see through the myths and on whom these mind tricks no longer work: vegans. These people are still very much in the minority (a sociologist would say their ideology is "not dominant"), and so it seems as if they, and not animal or dairy eaters, are the ones with an abnormal, irrational belief system. There are special restaurants for these "exceptions," and if they want to eat in a "normal" restaurant, they often have to ask the chef if they can have a meal prepared for them without using any animal products. If they want to eat on an airplane, they have to order a special meal in advance, one with "special dietary requirements due to health or belief," and if they eat at a friend or family member's house, it is very important to politely ask the host if they would be able to take their "special" diet into account. Occasionally this leads to stress on the part of the hosts because they are not used to cooking without using animal products. Frequently this leads to irritation, because the host considers the request to be strange or rude: Don't you want to try my amazing leg of lamb? Why can't you just conform?

In the next chapter, I will show you how and when vegans became normal. In the glory days of carnism, they are experiencing a mind-bogglingly fast change of image. Over the last few years, the vegan minority has transformed from a depressing group of unattractive pasty individuals into an elitist club of sexy six-packs, before expanding to include a huge number of tattooed, chubby, gay, and other "normal" people.

Intermezzo

We Didn't Know

There is a salty aroma in the kitchen, slightly reminiscent of clay. The cooking robot makes a beeping noise; the sous-vide machine is turning itself off. Winston Smith looks at the screen on the wall where, on a map of the city, he sees three red dots moving toward each other from different locations. His wife, son, and grandson are on their way home, and they'll soon be here. Perfect timing.

"Start rapid cool to lukewarm," he tells the cooking robot, which they bought and decided to call O'Brien for reasons long since forgotten. Winston puts his hands on his haunches, leans back, and straightens his shoulders.

"Initiating rapid cool," the device sounds through the kitchen—a friendly male voice, one that he now almost regards as one of the family. Winston claps his hands. "Music on, O'Brien. The cooking playlist."

Almost instantly the kitchen fills with the strains of a saxophone, and Winston starts whistling along. He runs his knife over the grinder a couple of times—he still does this the old-fashioned way, by hand, as he had been taught. This way you could get the knife as sharp as you want it, which an automatic sharpener just can't do.

"The dish must be cooled down to a lukewarm temperature," says O'Brien.

Winston removes a bag from the sous-vide machine and shakes out the contents onto a heavy wooden chopping board. A dark-red liquid runs into the grooves of the wood. He places the tip of the knife between his finger and thumb and slowly begins to make a cutting motion. For this recipe, precision is key: the thinner the slices, the better. The tips of his fingers are also red now. Good thing he's wearing an apron.

Suddenly the music gets quieter. Winston looks expectantly as his grandson Syme saunters in. Just behind him, hidden behind the tall figure of his grandson, is his son George. "Hi Dad," says George cheerfully.

Syme and George live in the same compound as Winston and Julia, in two detached buildings as part of the family home. Once a week they eat real food together as a family, most of the time in this very kitchen, and on those evenings it is almost always Winston who does the cooking. He has plenty of time to do it; his wife does not. Like most people of her generation who are still physically fit as pensioners, Julia works long days as a volunteer. To compensate for the previous damage done to the environment, she is very active in air and sea cleanup programs. Winston himself stopped last year, on his seventy-fifth birthday, after his back problems became too much to handle.

Winston would never say it aloud, but he is secretly extremely happy he now has more time at home. Just after he retired, he discovered how much he enjoyed going to the underwater farmers on the edge of the compound with Mr. Charrington, the robot dog, to do shopping, being able to cook for the family, and most of all being able to eat together at the dinner table on plates and with cutlery.

Unlike Julia, he has never been able to get used to the food pills and shakes. She thinks they're useful: "time-saving" she calls them. He finds the taste of these ready-to-eat meals as dull as the evenings without a family meal at the dinner table, evenings that seem to last even longer now that there is no need to spend time preparing and eating real food. The pills also make him gain weight, but Julia says that's impossible. In theory she is right, and he understands why. The pills contain exactly the amount of calories and nutrients to give you enough energy for the day without building up extra fat reserves. But in practice, they work differently. It seems as if his stomach doesn't want to remember that one pill can replace three old-fashioned mealtimes for a whole day. On the pill days, he always has major cravings, and it is only the fact that the refrigerator automatically tracks how much food it contains (and who uses it) that has prevented his apron from getting even tighter in recent years. The shakes, which they consume every other day, at least give him the feeling of something in his stomach, but he doesn't like the mushy texture.

"Great to see you, my sweets!" he greets his son and grandson. The electric kitchen door closes silently behind them. With his knife safely tucked behind his back, Winston bends down to hug Syme, but his grandson brushes him off, looking with disgust at the fingers on his grandfather's free hand.

"Gross, grandpa," he mumbles, as he takes off his rucksack, "you'll make my shirt dirty. I really don't get why you still make that finicky food." He goes over to the refrigerator and pours himself a glass of dark-green vitamin juice. Behind his back, George raises his eyebrows and rolls his eyes; Winston grins back at him. Neurologists on the news recently have claimed the part of the brain that controls empathy is much more developed among the younger generation than in his own carnist generation, but when Syme gets into his moods, Winston wonders if his grandson is perhaps the exception to the rule.

"Did you guys have a good day?" he asks. He leans up against the kitchen counter, looking from Syme to George and back again.

"Let's just say I had a productive day," George replies. His son is an organ grower at a huge health laboratory in the Eurasia region, only three hours away by high-speed solar bus. He looks happy, thinks Winston, but also a little tired. Not so surprising perhaps—he is fifty, after all, although he doesn't look it. The skin on his face is completely free of wrinkles, his body is thin with an athletic build, and his hair is full and dark.

"We've really been making progress with the donor lungs," says George as he takes off his waterproof shoes. Winston knows the lab is located in a floating compound, yet it is still strange to see his son walking around in his brightly colored water-resistant work clothes. On this side of the world, everything has stayed extremely dry (the compound buildings don't need to be flood-proof, only heat- and hurricane-proof). "We got the good news this morning. The results showed all the test humans reacted well to the tissue, so it looks like the operations planned for next week can go ahead."

"That's incredible!" cries Winston. He shakes his head. You can barely keep up with progress these days. All those people who died of lung cancer when he was young . . . soon it would be no more troublesome than a persistent cold. You'd feel pretty sick, but not enough to get a day off work.

"Sounds like a celebration is in order," he says. "Fancy a glass of white wine tonight? I've still got a very nice bottle around here somewhere." He doesn't wait for the answer before walking over to the wine rack, taking out a bottle, and placing it in the wine cooler.

Syme is noticeably quiet, but as Winston commands O'Brien to cool down the bottle to the ideal temperature, he also notices something wrong with George. His son looks at him as if he wants to warn him about something, but Winston cannot read his eyes. His son's shirt isn't giving off any stress signals—he must have turned off the function (what with George being a big fan of privacy and all).

Winston walks back over to the counter and continues preparing dinner. "What about you, Syme?" he asks, his eyes fixed on the knife in his hand. "How was school today?"

His grandson mumbles something that Winston cannot understand. He hears him jump off his board with a gentle plop onto the kitchen floor. "Put your solar board and your school things in the hall *now*, young man," says Winston, in a tone more patient than annoyed. "Otherwise Grandma will trip over it when she comes in." He acts as if he doesn't hear Syme's audible sigh. "What time do you have air football tonight, by the way? I can have the food ready in thirty minutes."

"Thirty-four minutes," the cooking assistant corrects him.

"Sounds great, Dad." George's voice behind him sounds warm. Perhaps there is nothing wrong with Syme after all, and he was just exhausted after an intense day at work. "There's no rush either. Syme told me on the way here he isn't going to training tonight. He's not feeling so good."

"I feel great," Syme snaps, "and I didn't say anything about missing practice!"

Winston turns toward his grandson and sees how angrily Syme is staring at his father. "When did I say I was sick?" Syme asks, his voice trembling. His eyes shimmer, red patches appearing on his neck. "I just don't feel like eating this food! I'd sooner just take a pill!" Syme slumps down in a kitchen chair and furiously begins taking off his board shoes. Winston thinks his T-shirt looks a little darker than when he came in, but maybe he's just imagining. The parenting coach also says it's normal for older teenagers in the new generation to regularly emit darker colors. It's part of the transition to adulthood, he tells himself.

Winston washes his hands under the tap. "Are you sure?" he asks, making sure his voice sounds less anxious than he feels. "I'm making beach beets with sea spelt and nut cheese! You like that, don't you?" He doesn't say what he is thinking; it is important for families to regularly keep eating together, especially for families with members from the carnist and non-carnist generations, and even more so for family members with non-carnists who are at a sensitive age. Food pills may well be easy and healthy, but according to nutrition coaches and other experts, they do not replace the opportunity for bonding a traditional mealtime provides. Eating together at the dinner table, the slowness of using cutlery, the chewing . . . all these activities create time for deep conversations, as well as the exchange of emotional information.

He takes four plates out of the cupboard, forcing himself not to look over at Syme. Things hadn't gone wrong again, had they? No, he told himself. Maybe George and Syme had fallen out? Problems at school? A broken heart? Hopefully he'll come around. "Your blood stats this morning showed your manganese levels are a little low," he hears George telling Syme. He imagines how his son looks at his watch while saying those words, zooming in on Syme's stats with his finger and thumb. "So Grandpa is making the perfect dish for your practice tonight."

"Beach beets and sea grains is a recommended dish for family member Syme because they contain relatively high levels of nitrite and niacin," the cooking robot adds. "Nitrite improves mitochondrial efficiency, and this helps improve athletic performance. Niacin produces hormones that are linked to stress regulation and sexual reproduction."

"O'Brien, off!" shouts Syme.

Winston stops suddenly in the middle of the kitchen, the plates still in his hands. From the kitchen the melodious sounds of saxophone can still be heard.

His grandson's T-shirt is now almost completely black.

"OK, what's going on?" he asks. George nods in Syme's direction and silently begins putting away his son's board and bag.

"Syme?"

His grandson's voice is now so quiet he can barely understand him. "I just can't believe you and grandma killed animals and ate them."

So there *is* something wrong.

"*Had* them killed!" Winston's voice is louder than he wants, he can't help it. He forces himself to breathe deeply in and out. Goddammit. They've talked about this so much lately. He lays out the plates on the round dining table.

"That's a rather important difference, Syme."

Calm down now. Think first, then speak. Every word matters. In his head, the words of the nutrition coach, the family psychologist, and newsreaders all begin to blend. What was it that the experts said? This is a normal part of the intergenerational Protein Revolution, and this involves a critical attitude from younger family members toward former animal-eaters. It is important now to be honest, and patient. Don't brush anything under the rug—that just makes things worse—but at the same time continuously emphasize that, at the time, this behavior was more socially driven than individually. "That was a key difference for us back then," says Winston, after he calms down. "I really hope you can keep trying to look at it from that perspective." He tries to sound friendly, as he practiced in the family therapy sessions. From the drawer he takes out four forks and spoons, which he lays on either side of the plates. Napkins! He mustn't forget the napkins, or the wineglass (if someone fancies a glass at some point). He sits down in the chair opposite Syme, wiping his fingers on his piñatex apron. It doesn't help: his fingers still look red. "That's the way it was back then, darling. Back then we didn't know any better. Everyone did it."

"I know," whispers Syme. His face looks pinched, and like every time Winston looks at his grandson not through a camera screen or without a beauty filter, he notices the small red pimples on his forehead. "And last time we were all at the psychologist's, I really meant it when I said I understand your decisions, but today in history class we watched a really intense documentary about food customs in the past, and I couldn't stop thinking about it all day." Syme looks at Winston pleadingly. "So you ate cows, right? And drank their milk?"

Winston nods. He feels the muscles in his back tensing up. In recent years he had answered dozens of similar questions from his grandson, each time more reluctantly. Tonight he just wants to enjoy being together with his family, and the food he has made: he wants to celebrate the success of George's hard work for the benefit of future generations, and not delve back into the past. It's all history. That was then, long before the breeding ban on livestock was rolled out by the government, which caused livestock numbers to plummet by 70 percent—previously that 70 percent of livestock in the Netherlands had been bred for export. Long before the government had ended subsidies for milk producers and started subsidizing vegetable growers instead, and long before the Protein Revolution began in earnest and television appeals were broadcast calling on people to stop eating animal products, in a last-ditch effort to slow down the effects of climate change.

He and Julia then switched completely to a plant-based diet, just like most other inhabitants of countries where plant-based alternatives were available. They didn't have much choice either. After government subsidies for meat and dairy were scrapped, prices rose dramatically. For most people, meat became as unaffordable as it had been in previous centuries, in the pre-factory farming age. These products then earned a bad reputation. The only places you could still eat meat were in special meat pods; small windowless rooms where animal-eaters who could afford the now outrageously expensive meat ate it out of sight of the new generations of children, so as not to upset them.

Julia and Winston never had that urge. The meat addicts they saw in those capsules always looked so sad and unhealthy: you could see it in their eyes. In the beginning, they would occasionally still eat a cricket burger (you were still allowed to eat insects), but they later opted for lab-grown meat.

From the corner of his eye, Winston sees George looking at him. He notices a look of pity on his son's face. He knows what George is thinking about. Both of them have seen what can happen when grandparents are not completely honest about their carnist past. The family psychologist said it was good for Syme to confront his grandparents with his anger and confusion: children of the new generation who didn't were more likely to divorce themselves from their families.

For this reason he is lucky to have George, who always tries to mediate between him, Julia, and Syme. After the Protein Revolution, there were plenty of children who had less understanding for their parents. They were angry at them for the pollution, the climate, the animal suffering. Thousands of them had cut off ties with their carnist parents, and more and more grandchildren never wanted to see their grandparents again. They chose to live together in separate compounds, in other regions of the world as far away from those former animal-eaters as possible. Winston wants to avoid a similar situation with his own family, so he'll have the talk with Syme once again, tonight and in the future, as many times as necessary.

"In the last years before the Protein Revolution, Grandma and I had stopped using animal products every day for a long time," he starts. He briefly looks back at the city map on the wall. A red light is flashing on the edge of the compound. Julia will be chatting to the guards, so she'll be out there another ten minutes or so. He wishes his wife would come home on time just for once; she always works so late on her pollution ops that every time Syme comes home upset, he has to deal with it on his own. "And even when we were younger and still ate meat, we only ever bought organic," he tells Syme. He rests his elbows on the table and leans down toward Syme. "That made no difference, of course, but we just didn't know. We really believed organic farming and slaughter methods were less stressful than nonorganic ones, and less painful." He stares down at his fingertips, which are still beet red. He suddenly feels exhausted. They hadn't even started on the environment yet.

3

From Pasty and Peeved to Sexy as Fuck

In brief defense of the establishment during the carnist age, early vegans certainly weren't a club that you'd want to be a part of. To be perfectly honest, they were rather odd, and kind of irritating as well. Okay, they were right about a few things, so it wasn't because of that, but they just didn't package their message (or themselves, as a matter of fact) in a particularly attractive way. The rumors that Hitler was vegetarian or vegan also did very little to help their image.

The Earliest Vegans: Adam, Eve, and the Pythagoreans

Early humans lived off seeds, fruit, and plants, so they were actually the first vegans, although calling their diet a choice would be a bit of a stretch: they simply ate what was available, and plants were just easier to obtain than animals. Following periods of climate change that led to fewer plants growing, humans switched to a cruelty-filled diet without any qualms. They hunted animals all the way up to the Protein Revolution, and as far as I know, no cave paintings from that time have been found that indicate any signs of a guilty conscience.

According to the Bible, we also know that Adam and Eve had a plant-based diet. They were fruitarians, eating seeds and raw fruit, preferably that had already fallen off the tree. On the sixth day, God did a promo for a fruitarian diet: "Behold, I have given you every herb bearing seed, which is upon the face of all the earth, and every tree, in which is the fruit of a tree yielding seed; to you it shall be for meat," God said enticingly. But even this already very rigid fruitarian diet was limited further, for God then imposed on his first human creations yet another commandment: "Of every tree of the garden thou mayest freely eat. But of the tree of the knowledge of good and evil, thou shalt not eat of it: for in the day that thou eatest thereof thou shalt surely die." You may eat the fruit of any tree in the garden, except the tree that gives knowledge of what is good and what is bad. You must not eat the fruit of that tree; if you do, you will die the same day. Great: just when you can eat only fruit, more kinds of fruit get made forbidden.

Adam and Eve therefore ate no animal products (and no grains, or vegetables, or forbidden fruits, at least until Eve just couldn't resist anymore . . . but that's a story for another book), but just like premodern humans, they did not in fact choose their lifestyle: they were ordered to by God. Adam and Eve were therefore another example of involuntary, unconscious vegans.

It would take a few more generations before there were people who consciously chose not to eat meat and other animal products out of compassion for the animals. But in Ancient Greece, this was already happening. Yes, in Ancient Greece, the place where during the Olympic Games cattle and other animals were gladly sacrificed to Zeus, where Apollo was described as the god of hunters, and where rich landowners indulged in lavish meat dishes even on weekdays, there were numerous anti–animal suffering groups under the leadership of the influencers of the day: philosophers.

The Philosophy of Veganism

The philosopher Theophrastus felt that eating meat was unnatural because he believed that animals belonged to the same family as humans, and thus in his eyes and those of his followers, eating meat was a kind of cannibalism. Orpheus and Empedocles also abstained from eating meat for similar reasons, and the Pythagoreans were the best known and in fact the largest group of animal-loving refuseniks.

Pythagoras, a name you probably recognize from the mathematical theorem you had to learn way back in school, a theorem our guy came up with himself. What they didn't tell you about him, however, was that math wasn't his only field of interest.

I fear your teachers didn't tell you this because you'd immediately forget the theorem, because the rest of Pythagoras' story is a lot more interesting. Pythagoras lived at the end of the sixth century BCE, and during his life he was the leader of a philosophical school. Pythagoreans lived according to a philosophy that forbade them from killing living beings and that ordered them to stay away from spilling blood and animal sacrifices. The reason behind such strict maxims was that Pythagoras believed that animals had a soul, and therefore killing them was cruel. He himself used far more elegant words to describe this:

> *Alas, what wickedness to swallow flesh into our own flesh, to fatten our greedy bodies by cramming in other bodies, to have one living creature fed by the death of another! In the midst of such wealth as earth, the best of mothers, provides, nothing forsooth satisfies you, but to behave like the Cyclopes, inflicting sorry wounds with cruel teeth! You cannot appease the hungry cravings of your wicked, gluttonous stomachs except by destroying some other life.*

Pythagoras additionally believed that if people got used to killing animals, then they would find it much easier to kill other humans.

In other words, a cruel diet leads to a cruel life.

How many followers Pythagoras had during his life has not been recorded, but what we do know is that you had to be a diehard in order to be part of the club. His most fanatical followers lived with him, living austerely, and were not allowed any personal possessions. Other Pythagoreans lived in their own homes but became members of the school he had set up, and tried to live according to his philosophy.

Math was not the only part of this all-encompassing philosophy, but it was certainly an important part of it: Pythagoras and his followers were convinced that numbers, numerical ratios, and spatial forms were the keys to understanding the world and reality. On their way to finding these keys, they enjoyed a diet of bread, raw or cooked plants, occasionally some honey, and very rarely some fish. Wine was also forbidden, for reasons that are not entirely clear, as were beans, the story being that Pythagoras believed part of the soul was lost every time you broke wind.

True or not, non-Pythagoreans at the time thought the philosopher was a little loopy; he was persecuted and forced to flee with his followers. Very little was heard from the Pythagoreans after this, but his teachings were never forgotten.

Sleeping on Newspapers

Guess what happened? Many centuries later, in 1847, a group of stubborn Pythagoreans formed a Vegetarian Society, this time in the United Kingdom.

This was a logical place to start a Vegetarian Society: the British were already familiar with India's mostly vegetarian cuisine, as well as with the ideas of Charles Darwin, who during that same period was preaching that humans and animals were not so different from one another, something contrary to what was believed at the time.

The word "vegetarian" was coined by these latter-day Pythagoreans from the Latin word "*vegetus*," meaning "one who lives a healthy life." The word "vegan" was created in 1944 by

Donald Watson and his wife, Dorothy, as a marker of a stricter variety of vegetarianism. They took the first and last syllables of the word "vegetarian," and the name for their more orthodox branch was born. Soon other similar organizations were established in the UK as well as in other countries: in 1850, vegan and vegetarian societies were established in the United States, in Australia in 1886, and in Germany in 1892.

These groups were far from popular, however.

This isn't such a surprise, as the vegetarian food eaten during that time was, according to science journalist Marta Zaraska, "boring and cooked to mush." In her book *Meathooked*, in which she examines meat consumption as a cultural norm, she describes what you would have been served if you had visited a vegetarian restaurant at some point between 1890 and 1920 as "pale, languorous carrots, with pecks of weepy boiled beets," all without salt or spices, as these, like alcohol, were considered harmful to your health.

How delicious.

Turns out these nineteenth-century vegetarians and vegans were almost as minimalist as the Pythagoreans of old. One group of vegetarians in the United States, the Grahamites, forbade everything that made life during that age a little more exciting: meat, tobacco, alcohol, and sex. The group's leader, Sylvester Graham, believed only eating whole-grain bread to be the best thing. Graham crackers, which are sold in American supermarkets to this day, are made with flour developed by Graham himself. His other great gift to humanity was the myth that you can go deaf from masturbating too much. (Gee, thanks, old-timer.)

John Harvey Kellogg, a leader of another group of vegetarians during the nineteenth century, believed that women should be circumcised and sleep on a mattress made of newspapers. At "Fruitlands," the communal farm established in Massachusetts by philosopher Amos Bronson Alcott, residents were allowed to wash themselves only in cold water, were forbidden from drinking tea, coffee, or alcohol, were not allowed to use any artificial light, and could wear only linen clothes. Wool was not allowed because it came from sheep, and cotton because it was made with slave labor. After seven months, this utopian community had collapsed (turns out it's quite difficult to run a farm with a group of individuals with limited agricultural experience). Yet this was much better than the vegetarian residents of Octagon City, Kansas, a community that lasted as long as it took for the snakes, mosquitoes, and Indians to drive them out. Finally, Leo Tolstoy, better known today as the author of *Anna Karenina* than as the vegetarian activist he was back in his day, called for the rich to give up their land and live in a more modest, more environmentally friendly manner. There are better PR tactics out there, let's say.

Alcott died aged sixty, Graham at fifty-seven, and the president of the Vegetarian Society at forty-eight. We know that they did not die because of a lack of meat, but rather of tuberculosis and other common illnesses of the age; but try telling that to the people living back then who had been shaking their heads dismissively at these stern, linen-clad, vegetable-eating weirdos for years.

The two world wars caused the popularity of a plant-based lifestyle to decline even further: if you were in the army, you were issued a meat ration and your options were to eat it or die of hunger. Priority during wartime therefore shifted from animals to humans. "Back then, nobody wept for the horses who died," writes Jojo Moyes in her moving book *The Girl You Left Behind* about World War I. Nor did they cry for the pigs: in the same book, the starving main character dreams of crispy roasted crackling, and the grease dripping down her chin.

Then we come to Adolf Hitler, and the persistent rumors that he was vegetarian or even vegan. This wasn't actually true, as according to his biographers and interviews with his chef, we know he rather enjoyed stuffed pigeon and other meat. The reason why this myth has kept circulating decades after his death was in part due to the Nazi regime being responsible for implementing a number of animal protection laws, and in part because Hitler and his buddies promoted a vegetarian lifestyle, as it would make Aryans big and strong. It may also be because in his final years he only ever ate mashed potatoes and vegetable broth, as is claimed, but this was not because he was a particularly ideologically motivated vegetarian, but more because he hoped that a plant-based diet would reduce his intestinal issues. Of all the worries he must have had, his flatulence and constipation must have paled in comparison.

Rage Against the Machine

In the twenty-first century, the number of vegetarians and vegans began to increase again. The world had recovered from the world wars, and the worst poverty and hunger had been banished, partly because of factory farming making the mass production of meat possible. But that same industry also began to provoke revulsion among more and more people.

Social scientists describe the postindustrial revival of the plant-eaters as a modern social movement. According to Alberto Melucci, one of the leading researchers in this field, a social movement is "a form of collective action, based on solidarity [and] breaking the limits of the system in which action occurs."

You might also say: the collective feeling of a social movement exists by grace of the difference with the masses. The social movement of vegans was a countermovement, a counterculture. Countercultures always exist in opposition to a mainstream or dominant culture, which members of this countermovement largely disagree with, and vice versa. Members of the dominant culture look at members of the counterculture with irritation, fear, skepticism, or a mixture of all these negative feelings.

There is an important difference between social movements like veganism and a "normal" social group: in a social group, people feel connected to one another, such as your group of friends from college you still talk to and go out for lunch with four times a year. You are part of this group because you like the other people in it and share wonderful personal memories with them. This is different in a social movement. It may well be the case that you find other "members" of your group incredibly irritating. You don't have to be friends with them, and you don't have to meet up with them. What connects you to this group is not the personal choices of individuals in the group, but a collective identity. Vegans identified themselves as people who didn't want to use animal products. They shared this chosen lifestyle together with all other people in their counterculture, and this brought them together, while distancing them further and further from the dominant culture.

In the postindustrial age, "vegan baiting" was so prevalent that Dutch journalist Amarens Eggeraat devoted a serious article to it in the magazine *Vrij Nederland* in which she tried to side with vegans. "Why do we hate vegans so much?" she wrote, as well as claiming that "they can never do anything right."

For a long time it was not unsurprising that vegans were not very popular: their food was boring and limited in scope, their diet, lifestyle, and philosophy shunned societal norms, and their external appearance did not fit the fashions of the day in any shape or form. They were underfed, wore sandals, had beards, or were pallid. Not exactly the kind of description you'd find on an online dating profile.

They also didn't behave in a particularly swell way. As the author of the above-mentioned article observed, they were crazy about animals and spoke for a "remarkably long time about hummus," or they were very angry and aggressive. If a vegan appeared on the news, they were always carrying banners with angry messages and kept going on and on about how they were angry at factory farming, carnist consumers, and the world. Some of them made secret video recordings in slaughterhouses and distributed them so everyone could see how poorly animals were treated. Sometimes a fervent animal-eater would be so shocked by these gory images that they would spontaneously become vegan, but those instances were an exception. Most animal-eaters kept their eyes shut, and only opened them when they were shown a more pleasant video about animals (one with puppies and kittens usually did the trick).

Grating Carrots

My carnist mother's eyes were wide open, however. Back when she was a student, she had tried to stop eating meat for several months. It was trendy back then, she told me. Oh, and she also felt sad about eating animals. She first told me about this decision when she took me to a meat-free restaurant for my seventeenth birthday.

I was extremely excited about our trip, as I hadn't been vegetarian for very long and the restaurant was the first fully vegetarian establishment I had ever been to. When the waiter came over to our table, I did my best to ignore the fact he was wearing a white linen gown and walking around in bare feet, and in a fit of festivity ordered a carafe of red wine.

"In my day, we still didn't have any good meat substitutes," my mother told me as we were perusing the menu. On the orange tablecloth a candle was burning that smelled of incense. "Nowadays you can buy veggie burgers at the supermarket, and look, they've got a tempeh satay, and even imitation steaks! Back during my veggie period, however, there was only tofu, and it was so tasteless that I didn't really feel like eating it every day."

Given that my mother had learned from her mother how to cook dishes where meat was the star of the show, as a vegetarian she had had to relearn how to cook meals that were as healthy as they were tasty and meat-free. This wasn't easy, my mother told me. "I can remember evenings where I spent hours and hours grating carrots because I had read in a vegetarian cookbook somewhere that the peelings could be turned into a sauce that would brighten up even the most tasteless of dishes. So every day your dad and I would eat meals drenched in carrot sauce." She looked at me with a pained expression. "One day we just couldn't bear to look at another carrot, so I think we just started eating less food in general, until after a few weeks we were so hungry and in such a terrible mood that we couldn't resist meat any longer."

Twenty years later, I could still understand why my parents made that choice. The food we were served at that vegetarian restaurant back in 1999 had no seasoning, the "wine" I had ordered turned out to contain no alcohol and was made from beetroot juice, and the main dish consisted of something with lentils, tofu, and lots and lots of carrot.

When you read this slightly musty, unpopular history of veganism, it seems almost inconceivable that this same social movement would, in the twenty-first century, become the place to be for all the cool kids. But this is exactly what happened. Within a few years, vegans would suddenly be seen as sexy as fuck, cool, and successful.

So, what happened?

Instagram, that's what.

Politics Through Pictures

In an article in *The Independent* in the spring of 2018, a journalist expressed her shock about the radical change in our image of vegans: *"How did veganism go from a mocked subculture to a mainstream lifestyle choice?"*

Her surprise stemmed from the fact that in that particular year, a plant-based lifestyle had become more popular than ever before.

#vegan was suddenly one of the most widely used hashtags on social media platforms such as Instagram, the term "veganism" was an increasingly popular search term on Google, and one by one a string of sports, film, and music stars publicly came out as vegan.

Formula 1 driver and five-time world champion Lewis Hamilton was racing faster than ever before on plant power. Music producer Moby began posting activist-esque posts about his vegan eating habits. Beyoncé claimed to be vegan, and Ariana Grande also became one. Kylie Jenner was in the club, Ellen DeGeneres was there for a while too, and Miley Cyrus got a tattoo that showed the whole world she was planning to become *Vegan for Life*: a V, the symbol of the Vegan Society. Score one for the Pythagoreans.

All these famous modern vegans have three things in common. The first is that, as opposed to their predecessors, they use a light, happy tone when talking about veganism. They show a large audience of people that, as a vegan, your life doesn't have to be a constant existential crisis about the mass slaughter of animals. They don't protest, they don't hold up banners, and they don't break into mink farms to free their "captives": instead they share cute photos of themselves cuddling up to a piglet or chowing down on a nut burger.

The second thing that this new generation of vegans has in common is that they don't eat animal products, but they're still *cool.* They prove that you don't have to be an outsider in order to be part of the vegan movement, and that an alternative lifestyle can fit in well with trendy fashion, pop music, mainstream films, and accepted beauty standards.

The third element that connects these new vegans is that they're all active on social media platforms. The vegan celebrities mentioned earlier are just a few of the ones who follow the animal-free diet, but at the time of writing, these seven vegans boasted a combined total of 564 million followers on Instagram, not to mention all the fans that follow them on Twitter, Facebook, Snapchat, YouTube, and LinkedIn. In my opinion, however, Instagram was perhaps the most influential of all of these for the Protein Revolution, as at the time it was the most visual of all popular communication platforms.

Instagram revolves around photos: photographic snapshots of a perfect reality. Photos of the lives users would like to lead if they were not so busy maintaining their Instagram feeds.

But Instagram also revolves around, and perhaps even more so on, the captions appearing underneath these photos, written by users themselves. First one or two sentences to support the image, then a couple of keywords to make the image easier to find on Instagram's search engine. When other people type this keyword into the search bar, they get an overview of all the images that match this tag.

Vegans are disproportionately active on social media, and it is thanks to Instagram that hundreds, if not thousands, of people have been converted.

Sexy and Vegan AF

True story: photos of colorful plant-based meals have an irresistible power of attraction for both vegan and carnist Instagram users. The carrots that my mother grumbled about back in the day now stand out with their bright orange color, and even the tofu she despised so much has been made to look incredibly delicious (especially when compared to the blood-tinged reddish-beige of a roast beef).

The hashtags that vegan Instagram users add under their photos turn the site into a free yet extremely effective PR tool. On Instagram vegans have found an easy way of dispelling decades of prejudice toward their philosophy and lifestyle. Under every photo of a beautiful meal you can find a political message. Shots of fuchsia-pink coconut smoothies are accompanied by messages such as #veganfortheanimals, #veganforyourhealth, and #veganfortheplanet, and rainbow-colored Buddha Bowls by #eatplantsbehealthy, #crueltyfree, and #theveganmovement. They are mini lines in the sand about justice, empathy, compassion, and what is wrong and right.

These same messages can also be found underneath photos that don't contain food, but in which vegans play a major role. Most of them are photos of them looking fit, slender, bronzed, ripped, and happy. These photos certainly do not fit the stereotypes from back when my mother was experimenting with a plant-based diet. Unlike those pale, unfashionable, angry individuals in the past, these vegans are actually easy on the eye.

There's fitness-mad @badassvegan, whose photos alternate between his ripped biceps and shots of him posing next to a bearded goat; and the adorable @deliciouslyella, who mixes pictures of her homemade plant-based meals with photos of herself and her husband (also adorable) and her two dogs (more adorable still). Then you have Rastafarian bodybuilder Torre Washington, with his washboard stomach, who grew up vegan; the blogger Cath Kendall, wearing a piece of material around her round, sculpted butt cheeks declaring how her booty is completely vegan; and many other sexy, successful Instagram users who caption their photos with #veganAF, #simplyvegan, #thefutureisvegan, #vegans_of_instagram, #veganpower, or #itscooltobekind.

Unlike the followers of carnism, as members of a counterculture, vegans need a political agenda to promote. For while animal-eaters are largely unaware that their diet and lifestyle are a reflection of a certain system of beliefs with associated ideas (eating meat is natural, necessary, and normal), and thus do not feel the urge to promote this belief on social media, vegans consider their diet and lifestyle as expressions of a fundamentally different belief, and they want to express their contrarian ideas (eating meat is unnatural and unnecessary and should be considered abnormal) via all possible means, with Instagram seeming to be the most effective of these.

More and more followers have begun to imitate these vegan Instagram users. They make their meals at home, share photos of them, and caption them with the same hashtags, but they don't necessarily do it because they subscribe to vegan ideology, but rather because they want to become as popular as these vegans . . . and as fit, beautiful, and happy.

Anthropologist Rivke Jaffe carried out a study into green and sustainable consumerism in the twenty-first century, paying specific attention to the rising trend toward veganism. At the time she carried out her study in around 2014, this was considered "more a choice of lifestyle than of ideals," which was a huge difference with the decades before this, and thus an additional explanation for the success of the postindustrial vegan movement compared to the movement's failures in earlier periods. "In the 1970s, 1980s and 1990s, veganism was left-wing and contrarian," concluded Jaffe. "Anti-capitalist ideas translated into eating plant-based. This was later reversed, with people beginning to eat vegan because it was 'in,' or because it was healthy." Jaffe calls this "eco-chic" in her report: promoting ethical consumerism and an environmentally conscious image while indulging yourself at the same time.

Regardless of the motivations behind these new vegans, the hashtags they created told users about their new way of living and eating and created connections with other members of the vegan movement. They met people online who they shared "something" with: a way of eating, a desire to look a certain way, or the pursuit of curves in the right places on their body. They shared a group identity, and as they began to get to know other vegans in the group better, they began to hear more and more about the ideology behind it. A year on, they looked back on their carnist past and could no longer imagine ever living that way with that ideology as a guiding principle.

Go with the Flow, Go Vegan

It started with a few dozen vegan influencers, who gained thousands of followers, and in the end the masses also found it #cooltobekind. The rise of veganism followed the rules of all major social and economic transformations. In any process of change you can divide humans into roughly three groups: the front-runners (often called pioneers), the followers, and the latecomers.

At first, front-runners from a counterculture are not taken very seriously by members of the dominant culture. They are often mocked when they promote their innovative ideas, and labeled naive, weird, or even dangerous. This is because their vision contradicts the status quo, and this is considered threatening by the majority of people who adhere to the dominant ideology.

A famous story of how threatening front-runners can be is that of Galileo Galilei, the father of modern astronomy. His belief that the sun, rather than Earth, was at the center of the solar system led to charges of heresy—at a time when getting in trouble with the Catholic Church could literally cost you your life. His vision was considered heretical because it was in conflict with what almost everyone at the time considered to be true: that Earth, and therefore humanity, were at the center of the universe. Court cases followed, and in 1633, when Galileo was sixty-nine years old, he was forced to recant his belief to avoid a death sentence; even then his freedom of movement was severely restricted as he was placed under house arrest. It was only in 1992 that he was posthumously redeemed, with Pope John Paul II announcing, on behalf of the Catholic Church, that Galileo had been right all along.

Things don't always end so badly for the front-runners. In fact, they only need to attract a very small group of fans if they want to make their new idea a popular one. In 2011, researchers at the Rensselaer Polytechnic Institute found that you can quickly change a belief held by the majority in a group as long as 10 percent of the people in this group are fully convinced by another idea. According to the researchers, this could happen with as low a percentage as 3 percent. If 3 percent of the members of a group are sufficiently convinced of an idea and propagate it, "the idea will spread like wildfire."

One in ten (or fewer) committed front-runners: this is all you need to get a political, religious, or other new idea accepted. The reason that such a tipping point in opinion formation exists is because people don't enjoy having an unpopular opinion. Humans are social creatures, and while they enjoy being right, they enjoy having friends a lot more. If a small group of people therefore decides that an idea is so important that they don't mind being ridiculed or laughed at for a while, and this small group expands to the extent that 10 percent of the population are members of it, then a point will be reached where the opinion of this group suddenly becomes *cool*. After this, more followers will come.

Galileo Galilei had followers and fellow front-runners, of course, but they simply were not around long enough to reach the tipping point, and had neither the power nor the nerve to help him.

In the period that Rivke Jaffe carried out her study on vegans and other "green" consumers, this new concept had already attracted a significant number of followers. Jaffe calls them the "eco-yuppies"; other researchers would call them "followers." Followers are people who like the idea of change and find innovation exciting, but who are not yet willing or able to take the first steps toward this transformation. They want to be a part of this exclusive club of front-runners, but in order to really change something about their opinion or behavior, they first need to know that the risks of such a change are small. For example, an innovation must be socially accepted (they don't want to look crazy), or it needs to have a financial or other incentive, and a new way of eating has to be just as tasty and healthy as the old way of eating (followers don't want their well-being or comfort to suffer as a result). In the case of the vegan movement, followers could observe from a safe distance whether their idols stay just as fit, beautiful, and popular after their shift to veganism, and if this appeared to be the case, then they would try "it" too. Another group of followers—corporations—saw that veganism was becoming a lucrative business, and copied the looks, slogans, and hashtags used by the front-runners in order to increase the popularity of their brands catering to this lifestyle.

Bringing up the rear of the social transformation process are the latecomers. These are the people who do not like change. They're scared of it, or think it's too much hassle, and prefer the status quo. Latecomers largely believe in conventional ideas, and make their voices heard when others begin to meddle. They sometimes express this by demanding punishment for the front-runners: they should be fined, silenced, or even excluded from society (exiled, or confined away, like Galileo was), and if that doesn't work (if the front-runners already wield too much influence through popularity, or if they are protected from punishment by law), then they must be mocked and jeered at.

This was exactly what the latecomers did to the growing number of vegans in the twenty-first century, until the moment when they wiped back their tears of laughter and found there were no more animal-eaters left with whom they could mock all the vegans. Now there were vegan products in the supermarkets, and vegan restaurants and non-animal-derived clothes and footwear were everywhere, and once the great Protein Revolution had taken place, the latecomers were shocked to discover it was not the vegans but *they* who had become social outcasts. A new idea spread like wildfire, they missed out on the big bucks, they lagged behind and now no longer belonged. A subculture became culture; a counterculture, the mainstream.

4

Giraffes for the Rich, Vegetables for the Poor, and Milk for All

The first time I made vegan pancakes for breakfast, my then husband took only a couple of bites before pushing his plate away in disgust.

"Just put some more maple syrup on it," I insisted, "then it'll taste all right." But I couldn't stop him from putting his hand in front of his mouth and spitting it out.

I kept on chewing. Eating pancakes for breakfast together was our Saturday morning tradition, and I had no plans to derail our romantic ritual just because I had replaced the eggs and cow's milk in the recipe with plant-based alternatives. He was overexaggerating, I thought. Maybe I still needed to do some tweaking to find the right balance of baking powder, flour, and almond milk, or perhaps I needed to work on the visual aspect of my attempts (they were rather pale and a little crumbly). "Chestnut flour actually works better than all-purpose flour," I suggested. "They'd also be great with some soy quark and blueberries!"

My husband began making a sandwich in silence. I took another bite of my pancakes, which did taste a little . . . odd. Somewhere between dish soap and green olives, to be exact.

Yet a little later on I helped myself to a second pancake, and then a third, and while I continued working my way through my culinary mishaps, I did my best to keep looking pleased with myself and make regular satisfied noises. I wanted to prove to my other half that cooking without dairy was a good idea no matter what. Not only was it kinder to animals and the environment, but it was also easy to do, and above all attractive from a culinary perspective. This is in fact what I had been telling him over the course of the weeks prior: we could still eat everything we enjoyed and found tasty, but from now on we could do it without that feeling of guilt.

My stomach was weaker than my will, however: more than half the pancakes I cooked that morning ended up in the trash. They were basically inedible.

That same afternoon I recounted the pancake fiasco to an American friend who had been vegan for a much longer time. She didn't laugh in my face, as I expected, but listened to me with a serious expression. "Vegan pancakes are really difficult to make," she said when I finished the story. "All my attempts were a disaster, so I gave up. You should perhaps try a new weekend brunch ritual. Ever tried tofu scramble? It's *a-ma-zing*."

Except I didn't want tofu for breakfast.

I wanted pancakes.

In other words, I wanted to keep living the life I had lived up until that point, and this involved having pancakes on Saturday mornings. I wanted to hold on to my title of queen of the kitchen; I didn't want to fail at something as simple as making pancakes. In the past, my partner had adored what he called the intuitive way I prepared recipes—unlike him, I hardly used cookbooks, but rather cooked based on instinct with the ingredients I found in our kitchen cupboards. And it worked: my husband praised my vegetarian lasagnas; he loved my *spaghetti puttanesca*, with a sauce of tomatoes, olives, capers, and anchovies, as well as my homemade bread and cakes.

But now he didn't long for my pancakes, a dish that a child could make, and I now couldn't even manage that.

Learning to Cook Again

"I swear to you, vegan cooking really isn't more difficult than cooking with meat or dairy," said British chef Derek Sarno. I had told him over the phone about my failed experiments, and he did his best to reassure me. "You just need to learn how to cook differently, so at the start it might be helpful to follow some recipes. Your intuition no longer works. First you have to learn how to combine ingredients if you no longer want to use eggs as an emulsifier. Then you have to learn what gives dishes a hearty taste if you no longer want to sprinkle cheese on them. This all takes time."

But after that you're onto something special. Derek told me he had never in his life eaten such delicious food as he had since learning how to cook vegan well, and neither had his dinner guests.

I believed him right away. He spoke so tastefully about juicy beetroot burgers and airy coconut pancakes, spicy *chili sin carne* and nut cheese boards with plum chutney, I found it difficult to focus on the questions I had for my book. All I could think was: I'll have what he's having.

Giraffe Meat

As I diligently spent my first weeks as a vegan going through recipe apps and cookbooks, my mind often turned to the generations of amateur chefs before me. There had been people in the past who had needed or wanted to learn to cook with unknown ingredients, and no doubt had at times caused their fellow humans to gag. What we do and do not eat is constantly changing.

In the days of the Roman Empire, for example, European elites enjoyed eating the meat of exotic animals such as giraffes and ostriches, which had been specially brought to Rome to be eaten during slave and animal fights in the city's arenas. These huge pieces of roasted meat were enjoyed in the afternoons, after a frugal breakfast of olives and bread. A cookbook from the Middle Ages included recipes for leg of camel, brown bear meat, and flamingo soup. There was also a recipe for goose liver pâté, made from geese that were force-fed figs in order to give them engorged livers.

Poorer people could not taste all these exotic dishes. This largely came from the increasing spread of Christianity across Europe during the early medieval period, and the increasing power of the church. It was also a case of meat not being so highly regarded in Christianity (Adam and Eve themselves had been vegetarian after all). During the Middle Ages, the Catholic Church had at least 140 to 160 fast days on which it was forbidden to eat meat, and in many monasteries eating meat was not done at all (poultry and fish were still allowed), except when someone was very weak, in which case it was believed that eating a strong animal would also make the person eating it strong.

A cookbook written by Peter Scholier, published in 1612, recommends preparing nutritious meals without using meat or fish, as this is *"more pleasing, more lovely and much healthier than all that meat, and all those slimy fish."*

For most of human history, therefore, poor people and the religious faithful largely filled their bellies with rye bread, grains, legumes, root vegetables, onions, and other tubers. Fruits and vegetables were consumed more for medicinal purposes than as sustenance, mashed into pastes or boiled into potions. According to the theory of the four humors, which remained popular even beyond the Middle Ages, fruit and vegetables were best avoided because they had features that doctors and scholars of the age believed were not good for the human body.

Roof-Hares

In the centuries before 1500, meat was not a permanent part of most people's diets, not because of the Church banning it, but because in the preindustrial era, meat was too expensive for the vast majority of people. During that period, 80 percent of people ate a largely vegetarian diet of bread, milled grain, and legumes.

Historical statistics show a strong correlation between meat and wealth. For example, in 1810, North-Brabant was the poorest province in the Netherlands, and the province with the lowest levels of meat consumption, at 55 pounds per person annually. When poverty in Belgian cities increased during the first half of the nineteenth century, meat consumption decreased in Antwerp by 22 percent and in Ghent by 60 percent. Poor and working-class people rarely or never ate meat, and only when wealth increased across the Low Countries did eating meat become more common. In 1850, the average Dutch person consumed 59 pounds of meat annually; by 1930, this figure had gone up to 110 pounds. Severe shortages during World War II brought this number down considerably (during the war, the Dutch were forced to eat everything from tulip bulbs, sugar beets, and rosehips to *dakhazen* or "roof-hares"—a nickname for cats), but by the 1970s, the Dutch were eating 187 pounds of meat per person each year. This figure roughly remained the same up until the Protein Revolution of the twenty-first century, with the exception of a small dip in the 1990s due to various outbreaks of infectious diseases that spread through meat.

The Milk Trend

In the years following World War II, not only did meat consumption increase again, but drinking milk also became popular. The Dutch Milk Board, set up by the dairy industry in 1950, began producing rather effective advertising campaigns for milk, cheese, and butter. The first large milk campaign started in 1958 and would give the industry a huge boost. This was important for the Dutch economy, as in those postwar years the country was working hard to create a welfare state. It was also important for dairy farmers, who had received considerable government subsidies to produce more milk, and when they did this en masse, it turned out the Dutch were not drinking enough milk. There was a surplus—a "milk lake." The Milk Board therefore tasked an advertising agency to give milk a new, more popular status as a "modern, sporty drink in the eyes of both old and (especially) young," and the campaign had to be primarily aimed at children.

And so it came to be: that same year, all Dutch newspapers ran an advertisement in which Dutch celebrities of the time launched the "Milk Pioneers Brigade." Children who drank an extra glass of milk each morning could become members of the brigade and, in addition to an arm patch, would be rewarded with free access to attractions such as zoos and theme parks. The campaign was a huge success: within six months, 320,000 children had become members of the Brigade, and that number would eventually reach half a million.

A few years after this campaign, the Dutch Milk Board was promoting a new advertising campaign, this time on television. An animated figure called Joris Driepinter (Johnny Three-pinter) advised young viewers to drink three glasses of milk a day for its health advantages. He did this during the 7 p.m. ad slot, the time when his target audience, children between the ages of six and thirteen, would be sitting in front of their television screens in their pajamas.

Joris was making this all up, but children didn't know this at the time, of course, nor did their parents. The health claims made by the Milk Board were never proven, and some were even rebutted in later years, such as the claim that (particularly young) people needed milk to get enough calcium. A 2018 article on the popular Dutch public health site www.dokterdokter.nl stated, "While it is true that milk contains a lot of calcium, you certainly do not need to drink it to get enough calcium. You are better off getting calcium from green leafy vegetables, beans, (gluten-free) oatmeal, sesame seeds, almonds, chia seeds, linseed, quinoa, fish, and broccoli," as well as mentioning that "there are even studies that show milk is actually bad for your bones."

Some of these studies were actually published in the 1950s and 1960s, but the Milk Board took no notice of these and launched the *"Melk, de Witte Motor"* (Milk: The White Fuel) and *"Melk is goed voor elk"* (Milk is good for all) campaigns. These involved collaborations with pop musicians from the time, sponsoring music festivals and other advertising campaigns aimed explicitly at children. The ad campaigns once again proved to be rather profitable for the milk industry: according to a 2014 sector report, the dairy industry was "one of the largest and most important agricultural sectors in the Netherlands" with a "strong, internationally oriented business model" that over the years has been characterized by "greater processing efficiency at fewer and fewer sites." In other words, Dutch milk producers were producing more milk in increasingly more efficient ways.

More Cows, More Milk: Fewer Companies, Less Land

Although the consumption of milk and other dairy products has increased exponentially, the number of milk farms has been decreasing. Small milk producers have been decimated in competition with large, government-subsidized dairy farms.

Between 1980 and 2016, the number of dairy cattle businesses in the Netherlands fell from almost 50,000 to fewer than 18,000. This meant that every day, on average, four small dairy farmers sold their cows to larger milk companies. This led to an increase in operational scale and more intensive milk production, with the average number of milking cows per business increasing from thirty-eight in 1980 to ninety-seven in 2016. In 2017, the largest 100 milk producers had an average of around 500 cows, compared to 288 ten years earlier. This new generation of cows, compared to their predecessors, were also given a lot less land: a report from Statistics Netherlands showed that between 2007 and 2017, the number of dairy cows per hectare increased from 1.6 to 2.3.

It wasn't just the dairy sector that was transformed but also the cows themselves. They were specially bred to give more and more milk at a faster and faster rate. Only cows that produced the most milk were bred, and less productive breeds were phased out. Newborn calves had increasingly larger udders, and the number of breeds of cattle in the Netherlands plummeted. In 2018, 99 percent of the country's dairy stock consisted of Holstein-Friesians, an American breed known for its high milk output. That same breed also dominates other European countries, the UK, and the US. Other breeds of cow, such as the *Fries-Hollands* cow (the most common cow breed in the Netherlands before 1975), the *Lakenvelder*, and the *Blaarkop*, are hardly found anymore, and are now officially endangered breeds.

I once spent an afternoon on the farm of one of the few Dutch cheese producers who kept Jersey cows: light-brown, relatively small animals with soft, brown eyes, who strolled lazily after one another, out of the meadow and into the milking shed. The farmer patted their flanks as he hooked up the milking machine to their udders, pointing out which ones had just had a calf and which ones would soon be inseminated with bull sperm. As he rubbed the teats of their udders with a disinfectant cloth, he explained why he preferred Jerseys over Holstein-Friesians: not only were his cows ill less frequently than the "inbred" Holsteins of today, Jerseys were also an "efficient" cow, which gave rich milk, full of casein, a protein that makes milk suitable for making cheese, with relatively little feed. The cheese farmer looked at his Jerseys with a loving gaze: "I would never want to milk any other cows."

Pregnant Top Athletes

Not such a good idea, according to an article that won a 2017 farming journalism prize, which compared the economic profitability of Jerseys and Holsteins, and the findings were more damning of the former. "Keeping Jerseys for normal milk production is not more profitable," declared the writer, although he added that the breed "could be a financially interesting use of an old barn with narrow milking berths," or to attract customers from niche markets ("these small cows will give businesses a USP (unique selling proposition) and distinguishes the owner from others who use the 'standard cow,' the Holstein"). There was still the question, however, of whether tapping a niche market is financially worthwhile for the average dairy farmer. "When subtracting all costs from revenue, with Holsteins, you have €3274 left over, whilst with Jerseys this is just €2984. We can therefore conclude that Holsteins are a more profitable breed than Jerseys."

What's more, over time they have gotten more and more profitable, as cows are giving much more milk than a hundred years ago. In 1910, an individual cow produced around 660 gallons annually. A century later this was more than 2,100 gallons. Dutch farmers managed to achieve this huge increase by targeted breeding, new milking techniques, the use of special, protein-rich feeds such as grain, soy, or fishmeal, and impregnating cows.

Let me explain. A dairy cow has to be constantly pregnant in order to keep milk production as high as possible. This is because she gives milk only when she is expecting or just gave birth to a calf—no different from any other mammal, humans included. Most dairy cows are therefore artificially inseminated every year to ensure they are constantly pregnant. During pregnancy, a cow is milked for most of that period, and once the calf is born, it is almost immediately taken away so the milk can be saved for human consumption. Under natural conditions a calf would drink from its mother for between six and twelve months: now it gets fed formula or powdered milk until it is slaughtered (this is the case for male calves, which, as I explained in a previous chapter, are considered the "waste products" of the dairy industry, much like how male chicks are considered the "waste products" of the egg industry), sold to another farmer, or, if it's a female calf who will grow up to be a dairy cow, forced to start eating solid feed after five weeks. Three months after giving birth, a cow is once again impregnated and the whole cycle starts over again.

This way of working is highly cost efficient for dairy farmers, and therefore good for the dairy industry, but not so great for the cows. In his book *Understanding the Dairy Cow*, Professor John Webster compared the exertions of a dairy cow to that of a human who would spend six hours a day running, all while being pregnant. No wonder these pregnant top athletes get sick so often: the high production rate of dairy cows leads to all manner of illnesses and ailments, in particular painful hoof inflammations, udder infections, and fertility disorders. Cows are therefore given antibiotics and other medicines to ensure they do not get sick and are ready to go before the next round of being pregnant and giving milk. Over the course of her life, a cow will go through this cycle around five times. After this her milk production goes down and she is slaughtered. On average, a dairy cow lives to be five or six, while their natural life span is between eighteen and twenty, with some reaching as old as twenty-five.

As I got off the train in Amsterdam after my visit to the cheese farm, my eyes were drawn to a large advertising billboard. On it was a photo of two packs of vegan milk, soy and almond, in a luscious meadow, with the prediction *"Plants are the new cows."* This was an ad for the Dutch margarine brand Becel, which became fully plant-based in 2019. According to a spokesperson, the company board decided to do this because it is better for humans and for the planet, but it would also be a logical decision because this, just like dairy cows before it, is "more economically profitable."

Kitchen Maid's Sorrow

The story of the rise of cow's milk in the Netherlands clearly shows that what we like to eat or drink is determined more by marketing and connected to popular ideas about what is good, healthy, and normal than by our own ideas and needs.

Marketing research suggests there are a number of different factors that decide whether a product promoted by a company or government actually becomes popular.

The taste of a product is one of them: if the Dutch had been as horrified by the taste of cow's milk as my husband had been by my first vegan pancakes, dairy farming as an industry would have quickly gone bust.

The price of a product also plays a major role. Until the industrial age, meat was too expensive for most families, so despite it tasting good, it was eaten very rarely. Subsidies for dairy producers from the Dutch government meant that products such as milk, yogurt, and cheese remained relatively cheap, which helped them gain popularity.

Another factor that determines whether a food product is a hit is how easy or difficult that food is to eat or prepare. This is an important reason why certain vegetables have been "forgotten" over time. For example, up until World War I, salsify was widely eaten, but with its hard, sticky roots, they were a pain to clean and prepare. It didn't earn the name "kitchen maid's sorrow" in Dutch for nothing! Housewives during the first half of the twentieth century couldn't switch to the other, easier vegetables on the supermarket shelves quick enough.

The final factor is how trendy and hyped a food product is. After the discovery of the beautifully attractive tomato by sixteenth-century European colonists in South America, it soon became *the* hip ingredient in kitchens the world over. The fruit looked good and eye-catching, the taste was easy to handle, and it had a special exotic air about it. A status ingredient for any mealtime! The somewhat less beautiful kale, however, needed slightly more encouragement before it would become all the rage. In the Netherlands, this mildly bitter vegetable is only used in the dinnertime staple *stamppot stew*, where it is mixed with mashed potatoes, while in the United States hardly anyone knew what kale was at all. That was until Oberon Sinclair, owner of the My Young Auntie PR agency, was hired by the American Kale Association to create a kale trend, which she succeeded in doing.

Cutting-Edge Kale

Sinclair mobilized her extensive friendship network of chefs, food writers, and food stylists, who made sure kale began to appear on the menus of New York's trendiest restaurants and influential newspapers and magazines began dedicating articles to it. This PR whiz also organized a guerilla marketing campaign that saw people writing "now serving kale" on restaurant chalkboards. She released tote bags and shirts touting slogans such as "kale(ing) me softly" and "the queen of green," which were then copied by trend-savvy fashion designers, and so it came to be that in 2017 Beyoncé was snapped in a T-shirt with the letters K-A-L-E on it. That same year saw 262 babies born in the United States with the name Kale, McDonald's use kale as a main ingredient in one of their salads, and kale chips and crackers become available in supermarkets.

Fame can be forced.

Instagram followers can be bought. Pay a lot of money and you can appear as an "expert" in a documentary, and if you are rich and shameless enough, you can always find a production company willing to film your life and broadcast it on TV.

The American Kale Association knew how to make the once unknown kale famous and desirable by consciously creating hype. Another company did the same thing with poké bowls (in fact with all "bowls," which could just as easily be called "bowl filled with vegetables and proteins next to each other," but that's not such a snappy hashtag), and others did the same with açaí, cauliflower rice, and chia. The Dutch milk industry also did the same but with milk. These all followed the same criteria:

Trendy: can be arranged.

Taste: tasteless to delicious, not gross at all.

Price: low enough to be accessible to a wide range of customers, high enough to give it special status.

Easy to obtain and prepare: check.

Attractive: 100 percent Instagrammable.

You don't eat what you crave. You crave what is offered to you to eat in an attractive way.

Clashing with the Powers That Be

No company in the world creates a new food trend or hype with the aim of losing money on it. There are companies, however, that work on creating new food trends with objectives that go far beyond just financial profit.

Pat Brown was a professor of biochemistry at Stanford University. He was well known for his passion, but also for being a bit of a troublemaker. He regularly clashed with the powers that be, so much so that over the course of his career he managed to radically change the balance of power in the scientific world. In the 1990s, he brought attention to the fact that scientists were publishing more and more articles in privately owned academic journals, meaning you could only read these articles if you paid for them. Brown didn't like this, as first this made academic knowledge unaffordable (and therefore inaccessible) for less well-off scientists, and second because the findings of publicly funded studies were appearing in commercial publications. He and several colleagues therefore started a so-called Open Access project, which, among other things, included a digital library containing published research that was free to access for all scientists. Pat Brown gained plenty of enemies with this stunt, but also managed to garner plenty of fame. In 2002, he was appointed to the National Academy of Sciences (NAS) in the United States, something seen as a great honor in his field. Only the very best researchers are allowed in, and they function as pro bono advisors to the US government in areas of general research, engineering, and health.

Shocking

Not all of Pat Brown's advice was heeded, however. During his first years as member of the NAS, his primary concern was the approaching climate crisis, and he often heard his colleagues talking about the same alarming research findings: the use of animal meat and dairy products for human food was responsible for more greenhouse gas emissions than all planes, cars, ships, trains, and trucks in the world combined. Livestock farming used and polluted more water than any other technology ever had, and this industry also occupied almost half of all the world's land, be it for cattle and sheep grazing or for growing soy, grain, and other crops to be used as animal feed. If the population of the world continued to grow and we all wanted to eat animal proteins, we would very quickly run out of land (at that point in time, a fifth of the land occupied by livestock had been overgrazed and exhausted). Brown found these facts shocking, but what he found more shocking was that so little was being done about this.

It was 2011. Brown had been working in his field for nearly forty years and had spent ten years as part of the highest scientific body in the land. He was in his late fifties and realized his time (and patience) was running out. He thus decided to dedicate the last part of his career to "something" that would combat climate change. Back then he didn't know what this "something" should be, but after an eighteen-month sabbatical he had found it: what the world needed was a new, more climate-friendly food trend, and if no one else was going to do it, he would create it himself. Brown got some money together and rounded up a team of scientists to come work for him. In 2016, a trendy restaurant in the center of Manhattan served the very first Impossible Burger—a meat-free burger that could bleed like a piece of beef.

Impossible Meat

The secret ingredient in Pat Brown's burger is heme, an iron ion bound to the pigment porphyrin, which creates a blood-like taste. Heme molecules are present in all living beings, but Brown's team extracted it from soy roots and yeast. Political journalist Ezra Klein declared the burger's discovery "life-changing," and one of the world's most influential meat-eating food bloggers called the taste of this plant-based meat substitute "impressive."

A fantastic compliment for Brown, but what was even more impressive was chef David Chang's decision to serve the burger in his restaurants. Chang, a huge fan of pork, had gradually removed all vegetarian recipes from his menus, and in an interview let slip that he believed animals were put on Earth to serve humans and that he "didn't want to live in a world with only vegetarians." His customers, however, had an appetite for this plant-based burger everyone was talking about. Chang tried one and was hooked. He bought Brown's burgers in bulk and sold them in his restaurant with a hefty price tag. A year later, these bleeding plant burgers were available in more than 1,200 restaurants and supermarkets across 20 states.

Investors gladly piggybacked off Brown's food trend. In 2018, his team had received almost 400 million dollars in investments from companies such as Google Ventures, UBS, Sailing Capital, and Temasek Holdings, and from individual sponsors such as Li Ka-shing (one of Asia's richest businessmen) and Bill Gates.

Gates himself had also previously invested a great deal of money in Beyond Meat, another company that was actively trying to create a new climate- and animal-friendly food trend. Founder Ethan Brown shared not only his surname with Pat but also his concern for the climate and his hands-on approach.

One rainy afternoon he began looking into which animal products produced the firm structure typical of meat, and realized that all these things (lipids, minerals, amino acids, and water) were also in plants. He looked for investors, used their money to build a laboratory and hire a group of skilled biochemists, and in 2012 brought out the first Beyond Meat chicken pieces, which had the same structure as chicken and were perfect for stir fries. The protein in his product primarily came from peas, which he ground up into a powder and mixed with other plant-based ingredients, such as fats from plant-based oil. The result tasted so much like meat that now you can find it in the meat sections of supermarkets such as Whole Foods. Bill Gates tried some and not only did he immediately invest millions of dollars in the project, he also wrote on his blog that he believed that he hadn't simply tried an inventive meat substitute, but the "future of food."

Ambitions

Eric Schmidt agreed with him. Schmidt is the former CEO of Google, and when he was asked to name the six innovations that would vastly improve the lives of humans, plant meat was high up on the list, above self-driving cars and even watches that know when you're sick before you do. This is because plant meat, as he wrote in his report, can help answer two questions: How do we combat climate change, and how are we going to feed the 9 billion people who will populate the planet by the year 2050?

Mark Post, a Dutch scientist at the Medisch Centrum van Maastricht, thinks Petri dishes are the answer. While Brown and Brown had been busy deconstructing plants, he had spent years refining the technology to grow beef in Petri dishes using stem cells from the muscles of cattle. After a period of time, he took these stem cells out of the dish and placed the cultures into a huge 25,000-liter bioreactor in order to turn them into beef.

One of these reactors could provide 10,000 people with "clean meat"—real meat, but with less guilt attached—and Post's creation does not create clouds of greenhouse gas.

Fellow Dutchman Jaap Korteweg is also making animal- and climate-friendly meat. He experimented with lupin, a type of legume, and made meat from soybeans. There is a photo of Korteweg on his website wearing a white shirt with splashes of what looks like blood, but on closer inspection they are in fact orange splatters from the bunch of dismembered carrots in his hand. For a long time Korteweg was a farmer and a huge lover of the taste of meat. His guilt made him want to become a vegetarian, but he also didn't want to give up his favorite food. With a team of food developers and chefs he started experimenting and testing, and thus the Vegetarian Butcher was born. The company has since been taken over by Unilever, which added plant-based chicken, bacon, bologna, hot dogs, nuggets, teriyaki chicken, meatballs, and even fish-free prawns and tuna to the company's line of products. Korteweg's ambition? To become the biggest butcher in the world.

This is a lofty goal, but still less ambitious than those of his fellow plant meat colleagues. A couple of years after starting Beyond Meat, its founder declared he wanted to replace all meat in supermarkets with plant meat by 2035, and when asked by a journalist how important his Impossible Burger could be for a sustainable future for the planet, Pat Brown coolly replied that his plant meat technology would eliminate the need for humans to move to other planets (an idea that was seriously being considered by some people as a solution to Earth's problems). "Mars is a really sucky planet compared to Earth," said Brown. "No one should ever want to go to Mars. There's no air on Mars. And yet people are saying we've got to figure out a way to get to Mars so we can have a place to live when we've totally destroyed this planet. Well, the impact we're going to have makes it unnecessary to go to Mars by saving this planet and keeping it habitable."

Both my vegan friend and top chef Derek, who had both recommended that I use recipes in order to learn to make pancakes again, were right, by the way. My pancakes are now no longer a source of tension in my relationships, and they are by no means vomit-inducing. I make them using coconut or almond milk (which we just call "milk" on our shopping lists, as plant-based milks have become the standard for us now), bananas, chia seeds, or whatever else I have at home. Some days are better than others, but they're usually tasty. Or maybe I should quote my American friend and say "friggin' ah-mazing."

5

Wanted: Man (20–40), Sporty, Sexy, Vegan

@magicalangel123 wonders "where that 'veggie Romeo' might be?" On a dating site for vegans she asks her "ethical soul mate" to write to her if he "has a big heart for everything life has, and room for me too!"

Twenty-one-year-old Lukas, from Brazil, describes himself as easygoing, friendly, and creative. He likes hiking and cycling in the outdoors, music, and drawing. He also cares for animals, which is why he doesn't eat them, and his future partner "naturally" should not either, although he adds "that should go without saying."

@ethicalvegan has two adult children, two darling grandchildren, and is now on the lookout for a "vegan gentleman" to spend the rest of her life with.

@maneatplants "eats pussy—not animals."

Roanne van Voorst

Meet the Vegansexuals

"Vegansexual" is the official term for people who live a plant-based lifestyle and seek out romantic or sexual relationships only with other vegans. You find them in large numbers on vegan dating sites, at vegan meet-ups the world over, on platforms such as Veg Speed Date ("for when you want to meet like-minded people lightning-fast"), or the vegan dating app Veganific. Researcher Annie Potts from the University of Canterbury coined the term "vegansexual" in 2007 after carrying out a study of 157 vegans in New Zealand and discovering that the majority of them fell for people who also adhered to plant-based diets, rather than people who ate meat. A study in the UK produced similar results.

The most important reason why vegansexuals want a vegan partner is to share important principles with a fellow plant-eater, face things together, and act accordingly. For example, you both don't think it's OK for "waste products" from the milk industry (male calves) to be slaughtered, so you don't buy cow milk from supermarkets because this would mean financially contributing to these practices. For vegans this is logical, and for vegansexuals doing otherwise can be a relationship deal breaker.

For me personally, "I'm a vegan for ethical reasons, and you?" is a bit of a weak opening line, but for vegansexuals this is a perfectly normal first date question. What's the point in carrying on talking otherwise, no matter how beautiful her eyes are or how sexy his three-day beard is? For vegansexuals, the notion of not wanting to date people who eat animals is as obvious as a refugee activist not falling for an out-and-out xenophobe. People who have completely different views on life do not agree on things that are important to them, and a lack of understanding is not sexy. For vegansexuals, the way someone consumes—both for sustenance and otherwise—is a sign of the deeper values and character traits of a potential partner. The contents of someone's cupboards reveal whether someone has compassion for animals or is concerned about the environment. This also suggests your date is not self-centered and is ready to take responsibility for their behavior: major plusses for someone who you might want to raise a family with.

According to the vegansexuals, another advantage of having a vegan partner is practicality. A vegan partner wants to be able to eat the same things and go to the same restaurants. They also want to order contraceptives from a natural drugstore, otherwise there's a good chance the condoms will contain casein, a by-product of the dairy industry, and the contraceptive pills gelatin, made from pork and beef waste products. (You didn't know that? This is what vegansexuals mean.)

Another important thing: some vegansexuals find a vegan partner more attractive because they smell better than animal-eaters, or they find the idea of kissing someone who might "still have pieces of meat in between their teeth" repulsive, or they lose all desire at the mere thought of their date eating a cup of cow milk yogurt with their breakfast the morning after. If this all sounds ridiculous, try to imagine meeting someone who spent all day eating slices of human meat, or the meat of golden retriever puppies. Chew, chew, chew, swallow . . . then those lips, all for you.

Preach

Not all vegans have been so strict when it comes to love. Myself included. About four years before I started writing this book and turned vegan, I fell head over heels for a man who was a fanatical climber. He was already a vegetarian when we met, but also a huge fan of dairy and eggs, and stayed that way even when I decided to turn vegan. He was used to getting the proteins he needed to build his muscles from animal products, so I would regularly make him a pot of Greek yogurt with nuts for when he came home tired after training, just like he would prepare a bowl of coconut ice cream with plant-based cookie dough for me when we spent an evening in watching Netflix.

Happily ever after? Not always. During the years we spent together, we both found these different foods and lifestyles rather difficult at times. When I was doing research for this book and leafing through page after page of reports about the dairy cattle sector and the chicken industry, it was hard sometimes for me to see "eggs" or "Parmesan" written on our joint shopping list.

It felt even worse having to buy these products, as this essentially meant I too was financially contributing to the animal suffering I was now so firmly against. At the same time, I knew my partner liked these things, and I believed I shouldn't impose my views on him. He should make his own decisions in life, and my principles were not necessarily automatically better than his. In other areas of life, he lived much more consciously than I did: he avoided plastic packaging at the supermarket as much as possible, and regularly picked up trash in the park near our house (I personally am often too lazy, too distracted, or in too much of a rush to do that). So who was I to lecture him? A lack of understanding is certainly not sexy, but thinking you know best perhaps puts even more of a dampener on your libido.

I wanted to prevent every meal together from erupting into a heated debate. I am more than my writing, I thought, as I stood in the middle of the supermarket looking up an alternative to eggs to use in a savory tart; I am more than my research. I did not want to turn into an activist: I just wanted to be his partner, someone he could laugh, cook, and chat with without a care in the world, just like before I started this book.

But sometimes I couldn't contain myself, and suddenly our evening would be turned into a lecture about the future of food. If he mentioned we were (still) out of eggs, I would start to explain with almost hysterical enthusiasm that I had made a quiche with chickpea flour instead of eggs, and then, in what I thought was a subtle way, let slip that in the year before, a total of 627,511,800 animals were killed in Dutch slaughterhouses, the vast majority of which were chickens.

"Insane, isn't it?" I asked.

He blew out the candles and put the cork back in the bottle of wine he had just opened.

"Going to bed already?" I asked, stunned, and slightly offended. "But it's a fact . . . can't we just talk about it? I should be able to talk to you when something bothers me, we don't have to argue about it."

We both had to learn how to navigate through new argument triggers. For me, this meant not sharing the things I read about, listened to, and reflected on at times when we were eating at the dinner table together. Often I wouldn't share them at home at all. I would instead write about them in my notebook and then call a vegan friend to talk about it or go exercise in order to let off some steam. I then consciously switched to a relaxing evening with my partner. For him this meant getting used to a partner who was suddenly adapting all his favorite dishes and wanted to buy different, new products. A partner who sometimes would, after the umpteenth interview with a farmer or animal rights activist, come home pale and with a fake smile on her face. A partner who got so personally involved in the world's problems that she brought them home and into the kitchen.

Our kitchen.

Trouble in Paradise

Other couples ran into similar problems, and some could not solve them without help. Picture this: a vegan guy falls head over heels for an animal-eating girl, and while still infatuated, it seems as if their dietary differences can be overcome. That is until their fourth date, when she plunges her knife into a rare steak and he cannot take his eyes off the streaks of blood slowly spreading over her plate, or until she introduces him to her parents and he has to turn down her mother's homemade cream pie.

Ouch.

Relationship problems between vegans and animal-eaters

are not always inevitable—there are happy mixed couples the world over—but fights are a frequent issue, so heated and so common in fact that Melanie Joy has produced a self-help book for mixed couples, and numerous vegan bloggers have produced articles with tips. More on that later, but first let's delve a little deeper into these problems.

After that difficult initial stage, many vegans find it almost impossible to listen to their animal-eating partner talking with glee about the "delicious piece of lamb" they want to order at the restaurant they're going to that evening, let alone if the animal-eater says that as far as they're concerned, their future children should be allowed to eat meat and dairy products. The more the vegan thinks about the meat and dairy industries, the less they understand how their partner is not concerned about it, or how they can think about it in any other way. On the flip side, the animal-eater gets annoyed about every "pleasant" meal together erupting into an ethical discussion about the food on their plate, and they feel personally attacked when they spot the disgusted gaze of their vegan partner once again fixed on their milk moustache. "It's not a personal attack," says the plant-eater with a sigh. "I'm not saying *you* are a bad person, I'm just pointing out that your choice to drink that has certain consequences."

But of course it's personal.

Comfort Eating

There is no choice in life more personal than what you eat. Food, family, and memory are all inseparably linked to one another.

Eating is the past; eating is tradition; eating is celebration; eating is comfort.

What you find tasty does depend on your tastes, but it depends even more on the things you have experienced and felt during your life. The chicken soup your mom would make when you were sick. The soft-serve ice cream that dripped down your chin on a hot summer day at the beach. The *spaghetti alle vongole* your boyfriend made when you ate at his place for the first time. The cheap, sweet liqueur you drank as a teenager during a night out and which still burned your throat for hours after you had thrown it up into a toilet bowl.

It doesn't get more personal than food, but it doesn't get any less political either. Every item of food you buy is, in some way, a vote. A vote for the dairy industry or for the plant-based milk industry; a vote for the organic meat sector or for the kale sector; a vote for a sugar company, for producers of clean meat, or for a mom-and-pop bakery on your street. Food is an expression of your political and financial support for an organization, and most people in the West do this at least three times a day, consciously or not. And voting is both a privilege and a responsibility, something no one else can do for you. Just like eating.

What you eat is one of the few things a person can and has to decide entirely by themselves. As soon as you become an adult, nobody else decides what you put in your mouth, unless you live somewhere where you have no free will (which I hope not), or if you are being force-fed (which I certainly hope not, because this would make you the human equivalent of a Toulouse goose, which even the most enthusiastic animal-eaters agree have unbelievably shitty lives).

If you can choose what you eat, every meal is a political action, a decision to support certain producers and to boycott others. It is an investment in one food development or another. Every bite you take drives the economy and politics in a certain direction.

Show me your kitchen cupboards, and I will tell you what you believe in:

Farmers should get paid well for their coffee beans, even if it costs me more money.

Coffee is far too expensive; the cheapest brands are good enough for me.

Food is medicine.

Food is fuel.

Animals serve humans.

Animals must not suffer so I can eat.

I know how wasteful and exhausting our food system is.

I don't know, I just want to eat nice things.

STSS

Enough of politics, let's move on to the implications of our food choices on our immediate environment: our dining tables, our beds, and our households. The rapidly growing number of vegans in the world will inevitably lead to the increase in the number of mixed-diet couples (a plant-eater with an animal-eater), and therefore an increase in the likelihood of arguments at the dinner table. This was also predicted by Melanie Joy, a specialist in relationship problems in mixed vegan/non-vegan couples.

"There is often friction in a relationship if one person becomes or already is a vegan and the other continues to eat meat and dairy," she tells me during a Skype call in the spring of 2019. She seems a lot younger than fifty-five, the age listed on her Wikipedia page. Her dark hair is tied back in a ponytail, and she is wearing a dark-green short-sleeved T-shirt. "Sometimes this is because the vegan is explicitly critical of the eating habits of the animal-eater, other times because while the vegan may not say anything critical, the non-vegan still feels criticized. The fact you no longer eat the same thing together can feel like a rejection of what you once were, or of who you are and what you think is right and normal."

Another relationship problem that often occurs has to do with the emotional transformation of an individual who decides to become vegan. "Prior to making this decision, many vegans experience a personal crisis. They have read or seen things that make them realize how milk and eggs are produced and how much animal suffering this process involves: things that they didn't know before. They feel lied to by the meat and dairy sectors, their parents, their teachers, newspapers, and companies they bought products from. Nothing they once believed in makes sense anymore. As a result, you can feel very lost and anxious, and this can have an effect on your relationship with your partner."

In her book *Beyond Beliefs,* which deals with communication and relationships between vegans and non-vegans, Joy describes this crisis as "secondary traumatic stress syndrome," or STSS. People can suffer from STSS when they have been the indirect witness of violence. First responders, for example, can exhibit symptoms of secondary stress upon hearing traumatic stories from clients and are bothered by them outside of work. Joy claims that vegans can also experience STSS when they watch TV reports or online videos about animal suffering, which they simply cannot shake off.

I think back to those evenings when I had to make the effort to forget what I had read, seen, or heard that day. It wasn't so much trauma, but sometimes it was really difficult to not feel sad or concerned, and at one point it took days for me to shake off that unpleasant sensation. Yet I found it slightly inappropriate to compare my sympathy for animals with the pity someone in the emergency services must feel when they are handed a child abuse or a domestic abuse case. At the same time, however, violence is violence. Why should I feel embarrassed about feeling something when confronted with this?

"Carnist society is a highly violent society," Joy stresses. "It's just that people don't see this because they have learned since childhood that mistreating animals is normal, natural, logical, and even necessary, and because this violence is hidden behind closed doors by the sector that carries it out. More and more people are seeing through this façade, however. Once you start to delve deeper into what really happens to animals in the meat and dairy industries, you begin to spot the signs of the large-scale animal abuse that carnist society deems 'normal,' in the refrigerated sections of supermarkets, on the highway as you drive past a truck full of pigs packed in up against one another, and at home when your partner or mother is sitting peeling a boiled egg."

Except often you are the only one who suddenly views these practices as abnormal, illogical, and violent. The rest of the world keeps on turning, buying and eating as usual. As a result, a vegan sees their partner, family, and friends as participants in something they are convinced is immoral, and no matter how kind and friendly a vegan has found these people, they must now begin to view them in a different light. In a certain way these meat-eaters are "responsible." They are complicit in the unnecessary suffering of animals, and a vegan, armed with this new knowledge and its proof, cannot understand how their loved ones can continue to follow the doctrine of carnism.

"Do you know that expression that soldiers use?" Melanie asks me. "Once you've been to war, you can never go home? A similar thing can be said for vegans."

Asocial

I again find myself struggling to compare vegan testimonies on animal suffering with the trauma experienced by war veterans, but I understand the feeling Joy describes through this comparison: this idea of discovering something you can never unknow, and as a result everything in your life that you once found normal and acceptable becomes abnormal and unacceptable, including the behavior of the people you hold dear.

Where I once bought eggs without even batting an eyelid simply because they were on the shopping list, I now find it difficult because I am aware that so-called free range chickens suffer unbelievably during their short lives. I used to spoon Greek yogurt into a bowl without a second thought, but the last time I did this for my partner, a thought suddenly shot through my mind: Isn't it odd that people determine how often and when a cow can get pregnant and then artificially inseminate her by shoving a long metal tube inside her? Isn't it odd that humans determine if and for how long a calf is allowed to drink its mother's milk, or that human hands and machines pinch a cow's teats, squeeze out her milk, and sell it? Isn't it extremely odd how we play God over the vaginas, stomachs, and breasts of cows?

I never really ever thought about those things.

Many other vegans have gone through a similar change, as I know from conversations with friends of mine and strangers who are vegans.

One time, J, a friend of mine, called me up in distress because he had been invited by his neighbors to their annual summer barbecue and was really looking forward to it. "I really like these people, we chat about things, we look after each other's cats and plants when we go on vacation, etcetera, so I feel like I should accept their invitation. But they are big meat and fish fans, and after looking into these industries further, I can no longer understand how people can't care about what happened to the animals that have ended up on their barbecue. I don't want to kill the mood by talking about animal cruelty, but it means I have to spend that whole evening acting differently to how I feel, and that seems so alienating to me."

My vegan friend V almost split up with her boyfriend when he thought she was "too dramatic" after she again broke down in tears after watching a film made by an undercover activist at a chicken farm, which showed how chicks, thrashing and struggling, are ground up alive. "I wanted to be cheerful after watching a film like that," she says with a sigh, "but I just couldn't get it out of my mind. I likewise find it difficult to understand why he doesn't even want to see those images at all. He finds it annoying when I tell him about it because it's 'unpleasant.' And I agree with him. But that's because our food production system is unpleasant, and if I get sad when I let these objective facts sink in and affect me, what does that say about him? Is he in denial? Or is he indifferent? I don't know what's worse."

Political journalist Ezra Klein, host of the popular and highly acclaimed *Vox* podcast, called his transition to veganism "one of the most disorienting, radicalizing experiences" of his life. Not so much because he ate differently than the people around him, but because he began to look at other people differently. During meals with good friends he saw them cutting into slices of meat, shoveling cream and eggs into their mouths, and all he could think was "what you're doing is morally wrong." Becoming vegan in a carnist society therefore potentially poses a threat to relationships, be they friendships, family, or romantic.

In Therapy

"What can mixed-diet couples do to save their relationships?" I asked the psychologist. Perhaps I sound a little too keen delivering this question, as Melanie Joy looks at me with an expression that can only be described as sympathetic, and asks me if my partner is an animal-eater and if I am a plant-eater.

I nod, slightly cornered. "I know, it's difficult," she tells me, smiling: she had told me before that her current partner is vegan, just like her. What about her exes? What happened with them and how did they deal with it? She lowers her gaze, hesitates, and then utters those telling words: "It's . . . hard."

Here was her advice. The most important thing, Joy explains, is for the plant-eater and the animal-eater not to further alienate one another. "For both partners, the feeling of disconnect they experience seems to be the result of food. The vegan often tries to solve this problem by trying to convince the other to also become vegan." They think that eating the same things will also connect them and their partner emotionally. This has the opposite effect, however: "The non-vegan feels pressured, and this creates an even greater feeling of alienation."

Huh.

My not-so-subtle summary of the number of animals killed per year that I unleashed on my partner that evening in the kitchen appeared, on second thought, to be a manipulation technique that not only was ineffective (my partner didn't eat any fewer eggs as a result) but also created further alienation between us. Not only did I find his carnist diet difficult to tolerate, but he felt the same about my vegan preachiness.

"The real problem for mixed couples is not what is eaten at home, but in feeling seen, heard, and supported by your partner." If that isn't the case, you feel insecure in a relationship. This applies to all relationships, including a mixed vegan/non-vegan one. "Feeling insecure happens when your partner denies or dismisses the thing that upsets you, or when they talk about it as if it's something positive." Picture the scene: the animal-eater talking gleefully about the "delicious piece of lamb" on their plate, or a plant-eater's partner denying that farm animals experience pain or fear, either during the slaughtering process or during their lives. "In order to restore that emotional connection with a vegan, animal-eaters do not need to eat differently, but they do have to show they understand why vegans are sad and concerned." In this way an animal-eating partner becomes a "vegan ally."

Likewise, vegans have to understand that while someone who eats meat or dairy contributes to carnism, this doesn't simply make them a "perpetrator," nor does it make them cruel or indifferent. "Humans are complex beings," says Joy, "and they play different roles in life. A vegan activist might wear plant-based sneakers that have been made by child labor in China. Is that person a perpetrator or a hero?"

Both plant-eaters and animal-eaters also need to communicate what their personal boundaries are. Indicating how far you will and won't go helps you find compromises together and avoid arguments. Does a vegan find it okay to buy animal products for a carnist partner? What about using meat, dairy, or eggs in cooking? Or looking on as the other person eats these things? And for the non-vegan, do they want to learn how to cook plant-based? Cycle farther to a different supermarket where they have more vegan options? Or is it important for them to eat meat without feeling the accusatory gaze of their partner from across the dinner table?

There we are. A free hour of therapy in the bag. A good excuse for writing a book, right?

Cavemanhood

We need to talk about men. Or, rather, masculinity. The vegan movement is growing extraordinarily quickly, and this has caused disruption not only in the world of relationships but far beyond that, particularly in our perception of gender. This is not the same as biological gender: gender is about culture, not nature. It's about identity and expectations of behavior: it distinguishes sexes by assigning qualities to men and women.

What these qualities are varies from society to society. In Western countries this distinction has for centuries been as follows: to be tough and rational is masculine, and to be gentle and emotional is feminine. Math and leadership are things that men are naturally good at, while women are better at languages and care. Men always want sex; women always have a headache. And "real" men eat meat, while vegetables are for women, sissies, and rabbits.

Exaggerated? Completely.

Cliché? Absolutely.

Biologically unproven? Hell yeah.

Nonetheless, stereotypes like this have a huge effect on the behaviors and thoughts of boys, girls, men, women, gender-fluid people, and gender-neutral people.

In 2012, American scientists published various studies that showed that consumers connect the consumption of meat to masculinity. "To the strong, traditional, macho, bicep-flexing, all-American male, red meat is a strong, traditional, macho, bicep-flexing, all-American food," concluded the *Journal of Consumer Research*. Macho American men who have traditional ideas about what it truly means to be tough and "manly" think red meat is a typically American, manly, powerful foodstuff. Many women in fact agreed with the men who were surveyed: they themselves often had no issue with not eating meat, but men, they believed, "needed" meat and were "expected" to eat it. The idea that "real" men "had" to eat meat was also dominant in other Western countries for a long time. Preferably the meat of a large, powerful animal—better a piece of beef than a fiddly little chicken. The (subconscious) belief that underlies this preference is that the eater will inherit the strength of that animal, so that he becomes "as strong as an ox." Brown-leather apron on, leave the beard, and just get grilling, because "fire, glowing coals, a nice plume of smoke that announces a caveman meal, is what we [men] want."

Certainly . . . except according to social psychologist Hank Rothgerber, this image of the carnivorous "caveman" did not come from prehistory. In an article with the revealing title "Real Men Don't Eat (Vegetable) Quiche: Masculinity and the Justification of Meat Consumption," he explains that the stereotypical image of the male meat-eater is much more recent. It would come to be created by companies who hoped to earn money at a time when a centuries-old belief about what it meant to be a man was ridiculed . . . by a group of outraged women.

Manvertising

We are talking about the twentieth and twenty-first centuries. In the West during that period there was more and more feminist critique about the unequal way in which economies and households in most countries had been run up until that point, with men as "leaders" and decision-makers and women as their subordinates. Patriarchal power began to wane as a result. In 1765, the British jurist William Blackstone stated in one of his commentaries on English (and later American) law that "by marriage, the husband and wife are one person in law: that is, the very being or legal existence of the woman is suspended during the marriage, or at least is incorporated and consolidated into that of the husband." Or rather, a woman had no legal right of existence within a marriage: her life was literally dependent on the way her husband decided to treat her. No matter if she had an inheritance or not, or if she earned an income of her own, she now possessed nothing, and her husband possessed everything.

But times were a-changin'.

In 1919, women got the right to vote.

In 1970, feminists in the Netherlands were proclaiming their desire to decide what happened inside their own bodies.

In 1980, abortion was legalized in the Netherlands.

That same year, the Equal Treatment (Men and Women) Act was adopted into law, which stated women must earn the same as their male colleagues when working in the same position.

In 1991, nonconsensual sex within a marriage became a crime in the Netherlands.

(I'll say it again in case you thought I had made a typing mistake: nonconsensual sex within a marriage became a crime only in 1991.)

These legal changes not only turned our idea of what it meant to be born as a woman on its head; they also rattled the image of what it meant to be a man. Traditional masculinity became more and more difficult to promote now that men were surrounded by trouser-wearing, well-earning, politically active, "no"-saying women.

Help for these confused men came from an unexpected place: fast-food chains and car manufacturers. During this time, their television and billboard ads were aimed exclusively at men who were struggling with their masculine identity. In ads for companies such as Domino's Pizza, Burger King, General Motors, and McDonald's, time and time again it was suggested that you could spot a real man by the fact he ate a whole lot of meat, and any lost masculinity could be restored by just buying more.

Take a commercial for Del Taco that was broadcast on American TV, where a man tries to put together a piece of furniture using an IKEA-style do-it-yourself kit. It doesn't work (which is not manly), and so he is encouraged by an all-knowing voice to buy a new burrito, which is mostly beef. "It's the only burrito beefy enough to feed the beast." The implication? When you eat beef, you also become a "beast," and that's what a "real man" should be.

Or what about a popular commercial for a Hummer, produced by General Motors: in the video we see two young, pale-looking men in line at a store checkout. The cashier (a woman) begins scanning the first man's items: tofu, vegetables, something with "soy" written on it, and fruit juice. Then the second man puts his items on the belt: huge packs of meat and charcoal for the barbecue. He looks at the first man's tofu and laughs. The first man seems to be ashamed: he looks away and bows his head. His eyes are then drawn to a Hummer ad, and he rushes off to the nearest car dealership. Later we see him tearing down the highway in his newly bought Hummer as he takes a bite from a carrot. The text over the screen reads: "Restore the manhood." Or rather, eating vegetables and tofu is less manly than eating meat, and if you do eat those things, then you have to compensate for this by buying something else that makes you super-manly (in this case, a huge, expensive car).

There was also an ad in the Netherlands in 2014 for the condiment manufacturer Remia, where an actor is seen eating a vegetable kebab while filming a war film. Just before he sinks his teeth into it, he is pulled aside by none other than Sylvester Stallone himself, who can barely protect him against all the grenades and other things suddenly being fired at him. "If you want to fight like a tiger," Sylvester advises him, "don't eat like a bunny."

As I write out and analyze these ads, I find them hilarious, and I am not blind to the fact that these ads present an idea of masculinity that is so overexaggerated that it becomes absurd.

Yet the humor and cynicism of advertisers does not make this representation any less powerful. A sensitive young man might laugh at it, but watching a lot of ads like this will in fact make him less willing to opt for a veggie wrap than for a hamburger. Eating meat not only looks tough, but it is also what's "expected" of his gender.

Men's magazines further reinforced this image of men as meat-eating "beasts." In 2004, linguistics professor Arran Stibbe analyzed six editions of the magazine *Men's Health*, which at the time had a global circulation of 3.5 million copies and was easily recognizable by the tanned six-pack of a male model on its cover. Stibbe subsequently concluded that the magazine's articles and images linked meat, especially red meat, to masculinity. Ergo: several times it was explicitly stated that eating meat was necessary in order to strengthen your masculinity because, according to the magazine, it would make you more ripped, healthier, sexier, and stronger (oddly enough, the fact that eating meat products that contain high amounts of saturated fats, cholesterol, and trans fats, such as burgers, hot dogs, steak, and bacon, is associated with cardiovascular diseases and erectile dysfunction was nowhere to be seen).

Shifting Notions

Around 2018, the male image once again began to shift, this time not thanks to riled-up women but to men who no longer wanted to identify with "tigers" or other animals. They didn't want to eat animals, period.

At the forefront this time were cooks, who wanted to show

that plant-based dishes could be just as nutritious and strengthening as meat; bodybuilders and other athletes, who had built their muscles using plant-based proteins; and vegan celebrities who spoke out openly and critically on the image of the male meat-eater.

In the UK, Derek Sarno and his business partner created a new image of what "real" men eat with their cooking company Wicked Healthy: the recipes they share with hundreds of thousands of followers online and in their cookbook are extremely barbecue-able, with the added bonus of improving your chances of getting a six-pack, as plant-based recipes are often much leaner than traditional, "mannish meat." In *Man. Eat. Plant.*, male vegans discuss why they stopped eating animals, and the book contains recipes for dishes that come out just as "bloody," "smoky," and "juicy" as those that were seen as typically "manly" during the Carnist Age—except this time Man Fodder 2.0 is made from beets, tofu, and other plant-based products.

What these men are doing is taking the concept of "masculinity" and turning it on its head, shaking it, and prying it open until it better fits who they want to be: strong, healthy, *Men's Health* cover model–worthy, and not violent. They only choose aspects of the traditional idea of masculinity that they find attractive: barbecuing, nutritious hearty dishes, and six-packs. The rest have been scrapped.

Actor, designer, TV presenter, and vegan Daniel Kucan did something similar in his blog. He let his readers know he missed meat, "every dang day" as he put it himself, but he felt the resistance to that desire to be a show of strength, which made it masculine. "I don't understand," he wrote, "since when is it manly to do what's *easy* instead of what's *right*?"

His justified question made me think of the author Jonathan Safran Foer, who wrote in his book *Eating Animals*: "Two friends meet each other for lunch. One of them says he fancies a burger and orders one. The other also fancies a burger, but he then thinks there are more important things than what he suddenly feels like eating, so he orders something else. Which of the two of them is the sensitive one here?"

Both Safran Foer and Kucan do not completely reject the traditional idea of the "strong man," but they have just adjusted it slightly. They suggest that for a "real" man, mental strength is just as important as physical strength, if not more so. Doing what you know is morally correct, even if—or perhaps because—you find it difficult and have to put in a little work; being disciplined: *that* is manly. Otherwise you give in to something you find tasty, comfortable, or easy, while deep inside you know it doesn't sit well with your deeper values. This, according to Safran Foer and Kucan, is not powerful, but emotional or sentimental—two character traits traditionally linked to womanliness, the "fairer sex."

For those who find this all a little difficult getting used to, there's always vegan Dominick "Domz" Thompson. According to many of his 193,000 Instagram followers, he is the embodiment of manliness and has become enormously muscled just by eating "what elephants eat," particularly plants. As for all the heterosexual female vegansexuals out there: as far as I can tell, Domz is currently sharing his bed only with his little white dog, Scruff McFly.

6

Plant Overdose

Oreos are vegan.

Ritz Crackers are vegan.

Hershey's chocolate syrup is vegan.

Vodka is vegan.

Swedish Fish are vegan.

Tater tots are usually vegan.

Lay's Originals and BBQ potato chips are vegan.

So are Pringles.

White sugar is vegan, and crusty bread rolls from the supermarket are vegan, and apple strudel from the most famous supermarket chains in the Netherlands is vegan, as well as their ready-to-eat cinnamon buns.

I think you get my point. Eating vegan is not the equivalent of "eating healthy." Veganism is a lifestyle that is based on a moral choice, not a choice that has to be based on knowing about food and nutrition. It is also not a diet or weight loss solution. The connection many people make between the two is the result of veganism's clever marketing.

In the Western world, many people want to be thin, fit, and healthy (often in that order), which is difficult if you commute to an office every day by train, you're too tired to go to the gym after work, and at the work canteen and at the station you are constantly tempted by hyper-fatty, sickly sweet, yet oh-so-delicious comfort food. No surprise, therefore, that one in three of us is constantly trying to lose weight, and every year the diet industry makes huge profits as a result: alongside pointless weight loss drinks, pills, and ready meals, they peddle hope. And hope is important. Hope springs eternal.

"Eat this, do nothing, and it will all work out," the diet ads promise, and of course their customers know this is nonsense. They're overweight but they're not stupid. They fully understand they will only lose weight by burning more calories, and that this is a slow process requiring immense amounts of discipline, patience, and acceptance. Yet anyone who has ever despairingly looked at their jiggly reflection at the end of another day of dieting knows that hope is much more comforting than disappointment.

The manufacturers of vegan products gladly piggyback off this knowledge, just like vegans who want to spread their ideology. Together they have created a new multipurpose diet: the Vegan Life—for all your wants and desires.

"Don't buy expensive diet pills: eat vegan and lose weight!" proclaim their posters and social media posts. This is a half-truth: while research has shown a vegan diet can include foods that are less fatty and also provide more nutrients that protect you against illnesses, you can also add all the fat, salt, and sugar to vegan dishes you want. You can gain weight from it, you can drink yourself silly from it, and if you really try hard enough, vegan food can clog up your arteries and give you a heart attack.

Chocolate spread is vegan.

Chocolate sprinkles are vegan.

Heinz Tomato Ketchup? 100 percent vegan.

Coke is vegan.

Sour strawberry belts are vegan. Sour lemon candies are also vegan, and the same goes for Skittles and Starburst.

Fries are vegan.

Cornettos and Magnums are available in vegan versions; at the time of this writing, Ben & Jerry's sold nine vegan ice cream flavors, including "chocolate peanut butter ice cream with fudge brownie pieces and peanut cookie dough." Sounds tasty, but far from healthy.

Halvarine, a Dutch brand of margarine, is vegan and, according to columnist Sylvia Witteman, a "dirty, disgusting Frankenstein spread." By this she means that, just like other margarines, it isn't made with animal fats but with plant-based oils and water, which through a chemical process are made spreadable. They get hydrogenated, a process that creates trans fats, which are known to increase the risk of cardiovascular diseases. Margarine may have fewer calories than cow milk butter, but it isn't automatically better for your body as a result.

Speaking of dirty, disgusting Frankenstein spreads, research has shown that some ready-to-eat vegan products contain large amounts of saturated fats and flavorings, especially salt. So much so that one innocent-looking broccoli burger could contain a third of your maximum daily allowance of salt.

Over the course of my becoming vegan the last few years, this was the case not only with ready-to-eat meat replacements but also with vegan cheese, which was still very much in the early stages of development (a polite way of saying that the available products were often inedible). In the Netherlands, originally most vegan cheeses were not unhealthy and did not contain huge amounts of salt (European food policies are thankfully rather strict in this area). In the United States, however, many supermarkets sold what amounted to congealed Frankenstein phlegm.

In contrast, a small group of vegan cooks, from all over the world, began making cheese from nuts and grains and experimented with aging and fermentation techniques where they let fungal cultures and lactic acid bacteria in products such as soy yogurt do their thing, just like what happens with conventional cheese. These vegan cheeses were on sale in the American stores where I did my shopping and, from those that I tried, were very tasty. They were also made from healthy products and scored higher in terms of nutrients than animal cheeses. They were laborious and time-consuming to make, however, and as a result were relatively expensive. A larger group of vegan "cheesemakers" tried to make the "cheesemaking" process faster and cheaper. To do this, they sometimes used chemical processes and additives, which are not healthy or tasty at all, or they used coconut oil as a main ingredient, which is rich in saturated fats. Some big companies managed to keep prices relatively low while still only using products that promoted a healthy diet. Be that as it may, many of the faux cheeses I saw in supermarkets in Philadelphia looked as if they were made of plastic, and tasted like it too.

Does this make vegan food not healthy then?

It does. Well, about as healthy or unhealthy as non-vegan food is. It depends on how much you eat, what you eat, and how you prepare it. It can make you sick or healthy. It can act as medicine or poison. It can make you high as a kite. It can be your enemy or it can be your friend. But I have a riddle for you: How is it possible that most plant-eaters *are* indeed healthier than animal-eaters?

The Riddle of the Healthy Vegan

If you shift from a typical Western diet to a vegan diet, there are two things you know for certain: (1) you stop eating meat and other animal products, and (2) animal-eaters around you will sooner or later worry that you'll get sick or even possibly die because of some sort of nutrient deficiency.

During my first years at university, when I lived off hot dogs, Nutella sandwiches, cookies, Coca-Cola, and budget wine, no one around me ever asked if I was getting enough vitamins and minerals. Yet when I turned vegan at the age of thirty, I was constantly getting questions and well-meaning advice about my health. Was I getting enough calcium? Protein? Iron? Vitamin B? Should I be taking multivitamin tablets, just to be sure? Shouldn't I get some blood tests, once a month at least? "You should definitely have an enema every now and again too," the secretary at my work recommended, as it was "good for fighting the toxins" in my body, which, she said confidently, would usually be absorbed by dairy products.

This concern is striking, as the vast majority of studies show that vegans are, in general, healthier than non-vegans. Many studies concluded that vegan diets on average had more fiber, antioxidants, potassium, magnesium, folic acid, and vitamins A, C, and E than those containing animal products. The studies I am referring to were carried out on large groups of people in numerous countries, over longer periods of time, and using so-called randomized, controlled testing, the results of which are published in peer-reviewed academic publications: a whole bunch of intimidating words that mean these studies are widely recognized as trustworthy and valid. In this chapter, I have based my statements purely on these "good" studies and consciously excluded any studies carried out in a relatively haphazard way, or which did not clearly explain how they were carried out. I did the same for "health studies" financed by the meat or dairy industries, or by self-appointed dieticians, or, although less common, those carried out by people who work for animal rights or vegan organizations. All such studies were not helpful for writing an honest and clear chapter about health and veganism. They did, however, help me formulate the following piece of advice: if you ever want to switch careers, consider becoming a health researcher, or going into the diet industry. There's a whole lot of work (and money) in it.

Rats vs. Humans

Meanwhile, researchers agree that vegan diet patterns reduce the risk of prostate, breast, and bowel cancer. Firstly, this is because vegans don't eat any processed or preserved meats, or red meat, regular consumption of which, according to the World Health Organization, increases the risk of developing colon cancer.

A second explanation for the fact that a vegan diet lowers the risk of contracting certain forms of cancer is that vegans eat a lot of things that seem to have a preventative effect against cancer. In general they eat more legumes, fruits, and vegetables than non-vegans (they have to find some way of filling up all that extra room on their plates). No more fried eggs with breakfast, no glass of milk with your lunchtime sandwich, no pepperoni on your pizza, and no piece of meat with your potatoes. If a brand-new vegan swaps all these things for products that are largely unprocessed and "wholesome," their new diet will consist of grains, fruits, vegetables, beans, peas, nuts, and seeds, and research has indeed shown that eating at least seven portions of fresh fruit and vegetables a day significantly reduces the risk of dying from certain forms of cancer.

A third reason that ensures vegans develop cancer less quickly, and have a lower chance of dying from it if they do develop it, is that their diets generally contain more soy products, which can protect people against breast cancer.

(You may have read somewhere that soy can in fact cause breast cancer. This is a stubborn myth born after scientists discovered that rats that had been fed lots of soy were more likely to develop breast cancer. The story was run by various newspapers and made many people scared of soy products. Subsequent studies showed, however, that for humans the effect was in fact the opposite: two to three portions of soy products per day, such as tofu, decrease the chances of developing certain forms of breast cancer. These studies also showed that women diagnosed with breast cancer could have better chances of survival if they ate soy products than women who ate no soy. The enormous differences between studies using animal testing and later studies using humans were caused by the fact that rats metabolically process soy in a way different from humans. But that news didn't reach the tabloids, it seems.)

More good news for vegans: studies that compared vegans to vegetarians and animal-eaters have shown that vegans are up to 75 percent less likely to have high blood pressure and are up to 42 percent less likely to die from a heart-related illness. Scientists believe this is because, again, vegans eat more fresh fruit, vegetables, legumes, and fiber—foods that do not contain the trans fats that mess with our body so much. Various studies have also demonstrated that vegan diets are much more effective at lowering blood sugar levels, LDL cholesterol (considered a major risk factor in arterial diseases), and total cholesterol values. A vegan diet, together with a so-called Low Carb High Fat diet, in which you do eat animal products and lots of fats but little or no carbohydrates, scores exceptionally high in terms of stabilizing and lowering blood sugar levels. This is particularly good news for the heart: lower blood pressure, cholesterol, and blood sugar can reduce the risk of heart disease by up to 46 percent. The last piece of positive news for vegans is that as long as they eat unprocessed and whole food, they have a 50 to 78 percent lower risk of developing type 2 diabetes compared to meat-eaters.

Two Big Buts

Vegans, you may want to stop the *na-na-na-na-nas* at this point, as here come the two big "buts" to throw a wrench in the plant-based works.

Big "but" number 1: many of the positive results of these studies could partly be explained by the lifestyles of their participants, and not just by their diets. You'd be surprised: in many of the studies I mentioned above, the blood work of people with a vegan lifestyle was compared with that of animal-eaters and vegetarians. But what if a large number of the vegan participants, but not the other participants, had certain lifestyle customs also considered healthy for humans, such as regular exercise, or meditation, or not eating much sugar or drinking too much alcohol, or something else that we don't yet realize is good for us? And what if it's these habits, and not the amount of veggies on our plate, or a limited amount of animal-based proteins, which might explain why vegans are sick comparatively less often?

We will never know for certain. Let me restate this: the studies I have discussed above are reliable by scientific standards, and so it's reasonable to believe that the results are not complete nonsense. Yet at the same time they are too limited to determine something as complex as human health. Not because scientists aren't doing a good job, but simply because it is impossible to make a study so all-encompassing that it takes into account all the lifestyle habits and circumstances of its participants. It would also have to be a lifelong study, so health researchers can study the long-term effects of all old and more recent behaviors, and always starting with a base measurement, or rather a measurement of a person's blood work from before they turned vegan. But that would have been back when these researchers were still baby-faced freshmen at college, and their test subjects were living off cookies and hot dogs.

Strap in, vegans, now it's time for big "but" number 2: vitamin B12 deficiency. It seems some of the concern of people around me was well placed. As a vegan, you can suffer from a vitamin B12 deficiency, and that's not something you want. B12 is vital for metabolism in our bodily cells: it helps protect your brain cells and plays an important role in converting carbohydrates, fats, and proteins into energy. If you suffer from a deficiency for a long time, you will not only have physical problems but also run the risk of harming your long-term mental well-being.

According to popular wisdom, B12 is only found in animal products such as meat, fish, and dairy. This is why the standard health advice to vegans is to supplement their B12, in the form of a pill or by eating meat substitutes with added vitamin B. There is nothing wrong with this advice, but the popular wisdom isn't 100 percent accurate, as B12 is in fact also produced by certain bacteria and fungi. You can find them in sources of water, in the earth, and in animals, and therefore also in your own body. If you don't want to eat any animal products, in theory you could go to a lake with a water bottle in order to get your daily dose of vitamin B12. Add a couple of handfuls of mud and you're ready to go . . . except for the fact that you'll end up getting sick from all the other less good bacteria you've also just ingested.

We can't rely much on producing B12 ourselves either, as it seems B12 is tucked away in a part of the gut from which we can't properly absorb it. Humans are therefore reliant on external sources of B12. This can take the form of animal products, or a pool of mud, but according to nutritionists, B12 can also be found in plant-based products such as seaweeds and algae, as well as in foods that have been fermented: a process in which bacteria, fungus, or yeast converts substances in one product into another product by changing its acidity, taste, smell, and appearance. We use fermentation to brew Belgian beer, for example, and tempeh, a block of compacted, fermented soybeans that has 40 percent proteins and provides iron, calcium, magnesium, and lots of B vitamins, is also made this way. In theory, you should therefore be able get your daily dose of B12 from eating and drinking plant-based products, but once again this is more difficult in practice.

One problem, for example, is that our gut cannot convert the B12 found in plant-based products, such as seaweeds, into a more active version, the one that's useful to us. How much of this we need to eat in order to absorb enough active B12, and how well our digestive system absorbs it exactly, has yet to be convincingly analyzed.

In the case of beers that contain B12, we have an additional problem: bottle-conditioned beers, which have had extra yeast added to them at the end of the brewing process, do in fact contain a considerable amount of vitamin B12, but when drinking alcohol, humans use more vitamin B, and so this cancels itself out. What about tempeh then? Well, it depends. While traditionally produced tempeh, made in countries such as Indonesia, usually contains lots of B12, tempeh made in factories in the West doesn't always contain it. This seems to be because not enough B12-producing bacteria can be found in tempeh factories in this part of the world: it seems Western hygiene standards have a very minor downside for our health.

The advice from vegetarian organizations, the Dutch Veganism Association (*Nederlandse Vereniging voor Veganisme*, NVV), and doctors still stands: if you are vegan, or want to become a vegan, get your blood tested to check if your body is able to correctly absorb important vitamins and minerals from foods, and supplement your diet with a small dose of B12 several times a week. (It's complicated to explain, but basically the higher the amount of B12 in the supplement, the less of it is absorbed overall. The best thing is therefore to take a lower dose of B12 but more frequently, rather than the occasional large dose.)

Confusion in Greenland

I've been following this advice carefully for a couple of months now, more for safety's sake, and to ease the minds of the people around me rather than genuinely believing that I need it. I've actually had my blood tested several times during my life, and only once have I had a vitamin B12 deficiency: after returning from anthropological fieldwork I had been doing in Greenland, where I had survived . . . on meat, fish, and animal blood.

When I was carrying out research on the consequences of climate change on indigenous hunters in Greenland, I ate what was at hand (I temporarily gave up my vegetarianism, for health reasons and also so as not to alienate myself from the Greenlanders), and there wasn't much. I had just turned twenty and was carrying out the research on my own. The village where I stayed for six months was on a small island in the North Atlantic and had a population of eighty or so hunters and fishers of both sexes. I moved into a small wooden hut that had been empty for years because my neighbors felt it wasn't very well insulated—it was too cold. It was winter: the sun didn't come up for months, and it was always between 14 and -22°F.

Just like my Greenlandic neighbors, I ate very little during those months, and because of the cold, against which my body needed extra energy, I regularly suffered from hunger. They hadn't caught much that year, and the store cupboards were soon empty. Sometimes a bird was shot, every now and then someone would treat me to seal blood soup, and rarer still I would share some dried whale meat with the other villagers.

As the months went on, I trudged more and more wearily through the deep snow when I had to go out in the mornings to fill a jerry can with water from a hole in the ice. I couldn't lift my snowshoes high enough, causing me to trip and fall flat on my face into the snow more and more often. My ski jacket no longer kept me warm, not even indoors. Sometimes, when the wind battered the walls of my wooden hut and blew snow in through the cracks, I shivered uncontrollably. I thought about food constantly.

At the end of my fieldwork, I was exhausted and skinny, my hair was falling out, I was constantly light-headed and confused, and my mind was like a sieve. Blood tests showed I had a B12 deficiency. This was caused not by a lack of animal products, I learned from the doctor who gave me a B12 injection in my left butt cheek every month, but by a lack of food, period.

I would later learn from my own research that if you have a diet containing varied products, and enough of them, in principle you should get enough B12, partly from the bacteria and fungi that are found in your foods, and partly because of the B vitamins that appear in a wide variety of foods in small quantities. This is true of a diet including meat, but also for a vegetarian diet and even for a plant-based diet, as long as you eat enough and eat a variety of foods, paying extra attention to potential vitamin B12 bombs during your meals. In my case, I had not eaten enough in general . . . even while I was eating animal products bursting with B12.

What I also learned was that a B12 deficiency is a rare thing among vegans (even among that group of plant-eaters who stubbornly refuse to supplement), as well as both vegetarians and meat-eaters. It largely presents itself among people who take certain medicines or people with an eating disorder, as well as among people in hospitals or older people in care homes. These latter two groups often have little desire to eat (or they get served food that is so salty, fatty, or tasteless that any desire to eat they may have had disappears immediately), and as a result do not eat enough food. The following advice therefore applies to all these groups, as well as healthy plant-eaters: take supplements, or eat veggie burgers or other ready-to-eat food products that have added B12, as a deficiency is no fun at all. I should know . . .

Protein

Of all the questions that vegans get, number one on the list has to be: "So, where do you get your protein from?" There's even a meme circulating among plant-eaters online, with the title "How protein takes up my time as a vegan." Below the title is a pie chart and next to it a key: pink stands for "struggling to find enough of it," while red stands for "explaining to people that it is easy to get." The circle is completely red.

I do understand the frustrations of vegans in this regard, but I also understand why these other people are so concerned. Like B12, proteins are extremely important for our health. Proteins, or rather the amino acids they are made from, are the building blocks of our muscles, skin, enzymes, and hormones, and they play an essential role in all bodily tissues. A protein deficiency is also a major problem, therefore . . . *in developing countries where people don't have enough to eat*. Let me stress again: in the West, protein deficiencies are extremely rare, except among older people, the sick, and people who don't eat enough or have unbalanced eating habits. Most foods contain some amount of protein, and so in wealthy countries it doesn't take much effort to get enough of it (that is, if you don't have a diet consisting purely of cookies, carrots, and potato chips). The recommended daily allowance of protein for the average active person is 0.8 gram per kilogram (or 0.36 gram per pound) of body weight. According to researchers at the University of Massachusetts Lowell, it makes no difference if you get your proteins from animal or plant sources: both have the exact same effect on your health. But because plant-based proteins are not as easily absorbed by the body and score somewhat lower in terms of essential amino acids, just to be safe, vegans are recommended to eat slightly more protein per kilogram of body weight (0.9 gram). People who do lots of strength or endurance sports require even more protein: endurance athletes need at least 1.2 to 1.4 grams per kilogram of body weight, while for strength athletes this is 1.4 to 1.9 grams.

More protein if you do a lot of sports plus more protein if you are vegan equals even more protein if you are a sporty vegan. Yet this is not as difficult as is often thought. In recent months, for example, I have been exercising three or four times a week: I do bouldering (a strength and agility sport), indoor climbing (a combination strength-endurance sport), and fitness once a week (strength). I weigh around 120 pounds, and so even if I counted myself among the group of the most dedicated vegan sporty types, I would need a maximum of 99 grams of protein. This is pretty easy to obtain via a vegan diet, and I feel strong, fit, and healthy.

What do I eat to get my protein allowance then? That'll be legumes, which are enormously rich in proteins. For example, 100 grams of cooked lentils contains 10 grams of protein: toss a portion of chickpeas into your lunchtime salad, and that adds 12 grams; a handful of peanuts (technically not a nut but a legume), and you'll have another 13 grams of protein. After an intense training session, a vegan protein shake (made mostly of peas) can help your muscles recover. But again: proteins appear in almost all "normal" foods. Tempeh has a lot, actually, with around 20 grams per average portion. A slice of toasted brown bread has around 4 grams, a pita bread almost 9 grams.

A bowl of oatmeal contains around 9 grams of protein; a portion of black beans around 15 grams. Quinoa and other grains such as buckwheat contain around 9 grams per 100 grams, tofu 11 grams, and a handful of almonds has around 21 grams of protein. Eat away, I say, and before you know it, you'll be a #plantbeast.

Teeth Fit for Plants

In fact, it is highly logical that people live perfectly fine without animal products in their diets, particularly because we were never originally designed to eat meat. We know this from archaeological research into the earliest hominids, which examined the minute details of their teeth, which we can use to determine what they ate. The teeth of the upright-walking humanoid *Sahelanthropus*, for example, showed signs of having chewed on a lot of tough fiber-rich plants. They also ate seeds and nuts. The *Sahelanthropus* lived around 7 million years ago and looked more like an ape than a human.

Many generations later, humankind's predecessors still primarily ate plants. From studies of teeth belonging to *Australopithecus*, a hominid that lived in Africa around 3 to 4 million years ago and which looked like modern humans, we know they had a diet that resembles that of modern chimpanzees: leaves, plant roots, lots of fruits, flowers, occasionally some tree bark, and every now and then a handful of insects. Much later, *Homo erectus*, an ancestor of modern humans, ate a diet of starchy tubers, roots, bulbs, fruits, vegetables, and seeds. Their teeth also seemed to be highly suited to chewing on plants, and not for ripping meat from bones. And things haven't changed. Just take a look in the mirror with your mouth wide open and you'll see our teeth are still very similar to those of herbivores such as orangutans. We have wide molars and short, blunt canines.

Imagine you're walking down the street with a rumbling stomach after barely eating anything for weeks (because, for example, you've been busy doing fieldwork in Greenland, or were living in a retirement home being served meals by an uninspired chef). You suddenly see a dead cow lying on the ground. If you don't have a knife, or anything else sharp on you to cut through the animal's skin, you have yourself a slight problem. Your canines won't help you, as they are too stumpy. We also can't open our jaws as wide as many felines and other carnivores can. Even if you managed to pierce through the skin (with your nails or something), you wouldn't enjoy eating it so much: raw meat is tough, and difficult for us to digest.

Later humans went on to eat meat not thanks to, but in spite of, our herbivorous teeth, and were able to do so because they discovered weapons. Around 2.5 million years ago, hominids began using the stones they had previously reserved for cracking nuts to remove meat from bones. They also seemed to have used them to cave in an animal's skull from time to time. Later we began using spears, and hunting animals became easier. Later still we discovered the bow and arrows, the gun, and finally what might be considered humankind's most destructive weapon: the domestication of farm animals, which are bred, engineered, impregnated, gassed, electrocuted, and killed by us in their millions, just so we can eat them.

Evolutionary Help with Eating

It may have been the case that we occasionally ate meat before the discovery of weapons. This was certainly true of groups of humanoids that once lived in places with few edible plants and with lots of animals roaming around. These humanoids were omnivores, researchers now believe. It also seems perfectly plausible that a hominid that mostly ate plants would, occasionally, gobble down a small monkey or bird that had fallen out of a tree. After this feast, however, they would have experienced terrible stomach pain. Our gut was, and still is, not short and flat like that of a cat or other carnivores, but is relatively long and ribbed, ideal for digesting fiber-rich, lean products such as fruit and plants, less so for meat, which is made up of a large amount of fat.

But evolution had a couple of tricks up its sleeve.

During the time we were developing more and more effective weapons and began to eat differently, our bodies adapted to our new diet generation after generation. Early hominids had an intestine that was ideal for digesting fruits and plants. They had an excellent appendix, a little bag at the start of the large intestine, full of bacteria, which could help them digest all these plant fibers. Plants in particular are not easy to digest. This is because plants don't have bones but still want to stand upright in order to grow toward the light. To do this, every cell in a plant is surrounded by a cell wall, a hard membrane that allows cells to stack on top of one another, similar to a tower of blocks. In order to break down these hard cell walls, a plant-eater needs wide molars to grind them down, as well as a long intestine full of bacteria that can tackle these cell walls.

"If one of these early humans had begun eating as much meat as the average American does today," science journalist Marta Zaraska told me via a Skype interview, "they would in fact have suffered from severe cramps in their large intestine. They would have become extremely nauseous, with a swollen, painful stomach." They could even have died from it.

Research suggests our bodies were gradually forced to adapt to eating meat, and this happened because humans went through a slow transition from eating a primarily plant-based diet, to an intermediary diet of mostly seeds and nuts, and finally to a meat-rich diet. This shift took place not only because of weapons but also because of periods of climate change: in certain places it began to rain less, causing the number of fruit and plant species to decline rapidly. Seeds and nuts were still (very much) available, and these were low in fiber but rich in fats. If our early ancestors did start eating more and more of these, as scientists believe, this would have stimulated the growth of the small intestine (where the digestion of fatty substances takes place) as well as the shrinking of the appendix (where fiber is digested). This would have paved the way for our digestive system to be able to process meat.

Evolution didn't stop there, though. In fact, in the last several thousand years of human existence, we have evolved quicker and further in order to make the consumption of animal products even easier.

Milk Gene

Recent studies have shown that since the Agrarian Revolution, we have developed various new gene variations that help us eat the way we do. Some people today are born with gene variants that protect them against diabetes and help their bodies regulate blood sugar. Other people are born with extra *AMY1* genes, which help the digestion of starches. And in Northern Europe, we are almost all born with a gene that helps us digest lactose in milk as adults, while most other humans in the world become lactose intolerant after they become adults.

Lactose is a sugar that naturally appears in milk products—it's what makes milk taste a little sweet. In order to break down lactose, you need an enzyme called lactase, which is made in the walls of the small intestine. A healthy baby makes enough lactase to drink and digest their mother's milk. After the first few months, lactase production slowly begins to go down; a baby begins to grow teeth and move on to solid food, so they no longer need to drink milk. This is the same for other mammals: in nature, cow calves stop drinking their mother's milk after around nine months. If you are born as a calf on a cattle farm, however, you are forced to stop drinking your mother's milk after a few weeks, as this will then be sold to humans. And if you are a human child growing up in a society in which there is a lot of dairy farming, after a certain period of time you will move from your mother's teat to that of an animal. You get a glass of cow milk with lunch, goat cheese in your sandwich at lunch, and a cup of yogurt as a snack.

The consequence of our love of milk is that around only 5 percent of Dutch people are lactose intolerant, while the vast majority of adults in South America, Africa, and Asia cannot tolerate milk at all. For example, Japanese and Chinese people lose some 90 percent of their lactase capacity in the first three to four years of their lives: give them two large glasses of milk, and they would get stomach and intestinal cramps.

Historically all adult humans on the planet were lactose intolerant. Up until 7,500 years ago, there were no dairies at all, and the animals had not yet been overbred to the point where they would calmly stand still and let us milk them. We didn't drink milk, so our bodies did not produce a special enzyme that helped us digest it. Just like we never used to have genes that helped us break down meat . . . but now we do.

Meat Gene

Professor Caleb Finch, a Yale graduate who earned his PhD at Rockefeller University, is a well-known name in the field of aging, health, and illness. He wanted to understand why human beings can live much longer than most apes, even though both have relatively similar genes. After years of research, his answer was: apolipoprotein E, or, in layperson's terms, the *ApoE* gene. This is a gene that helps transport fatty substances, such as those found in meat, and additionally helps humans deal better with dangerous infections, inflammations, and high levels of cholesterol. According to Finch and other health researchers, this gene has helped us be able to eat meat.

A little bit of evolutionary magic, you might say. Except that this *ApoE* gene has a slight downside. There are three different variants, and each person is born with different versions of the *ApoE* gene: *E2*, *E3*, or *E4*. This last one, the *E4* gene, is called the "meat adaptation" gene because it emerged during the time when humans still hadn't discovered fire or refrigerators but already wanted to eat meat. The *E4* gene helped our ancestors eat raw or spoiled meat without constantly suffering from diarrhea. This same *E4* gene, however, also leads to faster aging, and those people that have it die at a relatively young age.

Evolution strikes back: a couple of thousand years later, humans were born with an adapted, improved version of the *E4* gene: the *E3* gene (this illogical numbering convention, Professor Finch explained to me, was to do with a publication sequence, so don't read too much into it). What we know about the *E3* gene is that not only does it help humans digest meat, but it also increases their life span.

E3 is also the most common gene in the world, while only 13 percent of people are born with the *E4* gene. These people have had bad luck, as their average life span is reduced by four years, and they have a higher risk of developing heart diseases and Alzheimer's. If it's any consolation, in a competition for eating spoiled meat with people with other gene variants, they would win hands down.

The fact that humans do have special genes that make it easier to digest meat does not, however, mean they *need* meat to survive and thrive. The same goes for dairy. We really enjoy the taste, and over the course of thousands of years our bodies have learned to digest it, but the fact that we *can* eat it does not mean we are *designed* to eat it. We're also great at digesting cotton candy, but we are not designed for eating it, and we also don't need it.

What we do need are essential nutrients, such as B12, proteins, and good fatty acids. A bad vegan diet can lead to deficiencies in important nutrients, as well as a lack of fatty acids, iron, calcium, iodine, or zinc; the same can be said of a bad carnist diet. We have to be able to get essential nutrients from our food in one way or another.

If, in the twenty-first century, you live in a remote village in Greenland, you have no other choice than to get your nutrients from seals and whales; if you live in Indonesia, you can go a long way with tempeh and tofu; and if you live in the West, you can get these from a combination of plants, grains, seeds, and B12 supplements.

If you want to live to be old and healthy, then this is the best advice I can give you: make sure you are born with an *E3* gene, don't use any liver-damaging medicines, eat a plant-based diet with lots of vegetables, fruits, nuts, seeds, whole grains, and other nutritional foodstuffs, exercise regularly, preferably in nature, never ever let yourself be put into a retirement home and fed pre-made meals, have a B12 supplement or a B12-infused meat replacement every now and then (but not too many Frankenstein-esque products)—and then you'll succeed. Otherwise . . . well, I guess you could always get an enema.

Intermezzo

A School Trip to the Slaughterhouse

"Is everybody ready, and are you all sure you want to come in?" the tour guide shouts. She is trying her best to talk over the excited chatter of the group of students in front of her. "Good, let's get started. Here you can see the cages where the pigs were kept until they were slaughtered." The group leader pushes a large metal door with both hands, which slowly opens. Behind it is a huge, rectangular room with a high ceiling. It's dark and chilly and smells damp, with a faint odor of what to Syme seems like disinfectant. This part of the slaughterhouse museum is lit up only by thin strips of light shining down from the high, small windows. A six-foot-wide path runs through the middle of the room, and on either side of it are hundreds of pens, consisting of metal bars and concrete partitions.

Slowly Syme and his classmates begin to shuffle along the path. Some of them talk to each other quietly, one of them wearing an oxygen mask. Syme opens the door of one of the pens, goes inside, and does a circle on his board on top of the iron grating. So this is where they were kept. Waiting for what was to come. This was when they were somewhere between four and seven months old, or whenever animal-eaters had decided they had reached sufficient weight.

The guide claps her hands. "Can you all come and stand around me for a second?" she says in a cheerful tone.

"So let me tell you about this place." She is a tall woman in her forties, with long blond hair all the way down her back. Syme wonders what her name is: she didn't even introduce herself. He has to do his best not to keep staring at her. "Hanging on the wall to your left you'll see a docking iron, which was used to cut the tails off the piglets without anesthetic. This one is a slightly newer model, from 2017, and you can go a little closer to get a better look but please don't touch it, as it's one of the last examples of a docking device that has survived. Most of them were destroyed by activists after the Protein Revolution, or by former pig farmers who did not want to be reminded of their old profession."

A couple of Syme's classmates go over to the side of the room and crowd around the strange device. Syme can see it well enough from where he is. The object consists of a green piece of plastic, which must have been the handle, and a steel triangle, i.e., the blade.

"Huh," says Jones, one of Syme's classmates, "but hadn't docking already been banned for a long time before then?"

"Yes," the guide says, "it had been banned in the European Union twenty-five years before. This was because they knew very well that docking without anesthetic was extremely painful. You're cutting straight through a piece of flesh containing lots of nerve endings."

Syme sets his clothes a little warmer via his watch. He doesn't know whether it's the guide's stories or the low temperature in this dark, dank space that have suddenly made him feel so cold.

"But almost all pig farmers invoked an exception clause," the guide continued, "and the carnist government turned a blind eye to it. Officially you could only amputate a piglet's tail if a farmer could find no other way of controlling tail biting . . . such as configuring the stalls in a different way, or laying down some straw that the pigs can nibble on. This didn't happen here in the Netherlands, however, and because pigs were kept in small concrete pens by the thousands, without any toys or other stimuli and without any straw, they constantly bit at each other's tails out of boredom. This caused infections, which cost the pig farmers money, because you could not sell the meat of a sick pig.

"And so this led to 99 percent of piglets having their tails docked."

Syme's mind drifts to the home movie the family would project of Grandma Julia with her house pig. Pig Brother, or Piggie as the animal was affectionately known, had come to live at his grandparents' home after being saved from a slaughterhouse. In those early years after the Protein Revolution, almost everybody took in rescued farm animals, and pigs were especially loved because they were social and people enjoyed teaching them tricks. Piggie could fetch a ball, give his "paw," and go for walks on a leash with his owners. He would also sleep in their bed from time to time, and sit with them on the sofa. Sometimes when Syme hears these stories about Piggie, part of him thinks what a shame it is you can no longer keep animals in your homes. It seems rather nice, having a pig or a cat as a companion. On the other hand, Mr. Charrington, the family's robot dog, knows lots more tricks than Piggie, requires very little care, and will never die (as long as you don't flip the "off" switch).

"What's that?" One of Syme's classmates leans back his head and points to the museum's ceiling, where every few yards, small metal nozzles hang from the ceiling. "Are those the gas showers?"

"No," replies the guide, "those were sprinklers, which sprayed water to clean the pigs before they were killed. Pigs weren't gassed, that only happened to male chicks, who themselves had their throats slit or were ground up in a mincer. Pigs and cows had their jugular vein cut by the animal-eaters after they had been stunned. If the stunning worked, that is. Otherwise they did it without."

The tour guide stares off into the distance behind Syme's head for a moment. She is silent for a couple of seconds, then shakes her head, almost invisibly, blinking a few times. When she starts talking again, she doesn't seem as animated as she was before: "If there are no more questions, let's move on to the conveyor belt where they hung the pigs so they would bleed out, and then you'll also be able to see the hot water tanks they were dipped in so the hair could be removed. You there, the boy with the red hair at the back, did you have a question?"

Syme sets his board a little higher, so his head sticks up over the rest of his classmates. "How many . . . ," his voice cracks, he clears his throat, ". . . were killed every year?"

She starts nodding as if she had been expecting the question already. "Chickens or pigs?"

"Both," replies Syme, unsure if he really wants to know. His cheeks are glowing, his heart pounding.

"In the Netherlands, the egg industry was using 30 million hens every single day. Remember that this country was once the biggest egg exporter in the world, so there was plenty of work for the slaughterhouses. On top of this we killed 1.5 million chickens every day for the meat industry—these were the broilers between six weeks and a year old, and the laying hens that were all laid out. As for the pigs . . . at the peak of carnism in the Netherlands, in 2018, around 16 million pigs were killed every year. Here, at this pig farm, the animal-eaters killed around 45,000 pigs per day. This occurred roughly when they were infants, so usually when the pigs were a couple of months old, depending on how quickly they reached the right weight." The guide explains the facts quickly, without hesitation, without stumbling over her words. She must do this at least three times a day, thinks Syme, for other school groups, corporate outings, and tourists. Rather her than him. She continues. "The slaughter date was not determined by the pig's age, but by their weight. They gained weight faster and faster due to the way they were bred. In 2013, the average weight of a pig at slaughter was 205 pounds; in 2017, it was 211 pounds." She smiles at Syme reassuringly. "The number of pigs here was relatively low: in the United States around that same time, it was 112 million per year."

The guide opens a door at the side of the pens and leads them into another, this time much brighter, room. At eye level, a metal scaffold runs along the walls, with a coat hanger–like device hanging down from it every foot and a half. "This was the real slaughterhouse," she continues as the students gather around her, "a factory consisting of two parts, although it's hard to imagine, with no workers here. In the past, where we are standing, there'd be people walking around here in blue overalls. Here the pigs were killed, bled out, and dehaired. Over there, at the back of the room, only people wearing white overalls were allowed. These were the people who worked on what was called the 'clean line.' This was where they sliced up the dead pigs into cuts." Syme looks at the floor, which is spotlessly clean. He's glad he came on his board today, though, and that the soles of his sneakers don't have to touch the floor.

The guide's finger makes a circular motion in the air. "See those hooks hanging from the 'clean line'? The pigs were hung from these by their back feet, so they'd bleed out onto the ground." She presses a red button on the wall, causing the conveyor belt to start moving with a loud hum. Syme has to move out of the way quickly to avoid a hook seemingly going straight for his face.

She shouts over the noise of the machines. "As you can imagine, this was hard work for the slaughterhouse employees. We now know that over time most of the employees became used to the blood and cries of the pigs—they became numb to it, and even began to find the work tedious. But not only was this mentally difficult work; it was also physically difficult. Six hundred and fifty pigs were killed and butchered here every hour, so it was a high-tempo operation. Lifting up a nearly 220-pound animal by its back feet and slitting its throat is not an easy task, especially if you have to work quickly. Sometimes this went wrong, and so the pig had to be killed by hand, otherwise it would thrash about on the hook until it had fully bled out. There are no official figures for how often they failed to kill a pig in one go, because the industry made sure that those kinds of figures weren't published, but from interviews with whistleblowers who worked in these slaughterhouses, we know around one percent would be a very conservative estimate."

That's more than six per hour, Syme quickly calculates in his head. Forty-five hundred per year. He suddenly begins to feel dizzy, reaching his hand out toward the wall for support, but quickly pulls it back. He doesn't want to touch anything, no matter how thoroughly the stonework had been cleaned at the end of the last shift. "The work was also extremely poorly paid. Many workers came from Poland or Cabo Verde, and worked long days for about €10 per hour, which is about 2 cryptos in today's money."

The guide turns off the conveyor belt. "We've now reached the end of the tour, unless anyone really wants to see the hot water bath where the carcasses were dipped? Or the cutting and packing room?" Syme looks around anxiously, but to his surprise, most of his classmates have already started heading toward the exit. "No? You've seen enough? Good, that leaves me with one last important question for all of you. Who among you has animal farmers or butchers in their family?"

Those who have already reached the exit stop in their tracks and turn around. Nobody says a word. Syme holds his breath and, from the corner he found himself in, glances to the side. There, barely six feet from him, is Parsons. Everyone knows Parsons's grandparents had once owned a chicken farm. His head is bowed and, without looking up, he raises his right hand.

"Thank you for sharing. And who among you had grandparents who were in the countermovement?"

From what Syme knew, the countermovement was a relatively small but powerful group of die-hard animal-eaters who had long tried to stop the Protein Revolution. Many of them worked in the meat and dairy industries, others worked for the Food and Consumer Goods Authority, or for carnist governments that maintained close ties with the animal foods sector. Together they had tried for years to stop the taxes on meat and dairy, and to obstruct the shift to a plant-based food economy in other ways. This led to a European Union Court of Justice ruling that soy milk and almond milk could no longer be called "milk" because these names could lead to confusion among consumers. Products with a dairy-related name, such as milk, butter, cheese, cream, and yogurt, could only be called so if they actually came from animals; the same applied to products with a meaty name, such as sausages. Years later, when Syme's mother was pregnant with him, the court reversed this decision completely: from that point on, plant-based products became the standard, and packaging had to state explicitly whether an animal had been killed in the making of a product.

The countermovement had also spread dozens of fake news reports claiming that certain famous vegans had died because of nutritional deficiencies caused by their plant-based diets.

They made a documentary that warned that, without the meat from farm animals, there would be worldwide food shortages and famine. A small group of radicals were even charged with committing acts of violence, such as the outraged American cattle farmers who had attacked the employees of a nut milk factory, and the terrorist group who in 2025 set off a bomb in the toilet of a popular McDonald's store soon after the company had announced it would be replacing beef patties with a plant-based alternative, with the only "meat" option being the McCricket.

"My grandpa was," says Jones, a girl standing close to Syme; another boy, closer to the exit, silently raises his hand.

"I think it's very important," says the guide in a friendly tone, "to emphasize that everyone here is blameless, and, to some extent, your grandparents were too. It was a different time, with different ideas and a different level of knowledge. People back then didn't know any better."

The students around Syme begin to murmur.

The guide straightens her back, raising her head slightly.

"My grandparents, the Goldstein family, owned this slaughterhouse, and voluntarily decided to stop when the Protein Revolution got under way," she says solemnly. "My father, Emmanuel, however, didn't agree with this, and as a young man played an important role in the countermovement."

The room is suddenly deadly quiet. Syme notices he has started wobbling on his board: his legs seem to have turned to jelly. He certainly did not expect this from her. Not that it makes any difference, of course: she is a different person from her family, and you can never tell by looking if someone is a meat-eater or a plant-eater. "When I was your age, I felt ashamed," she continues. "But I now know that my parents and their parents were no guiltier than all those people who cooked and ate the pigs they killed. The same for the chickens, cows, and goats. Do we all agree?"

"Yes," mumbles Syme, and around him his classmates agree with her. Syme sees Parsons rubbing his eyes.

"Excellent," says the guide. "I'd like to thank you all for your attention during the tour. We have online psychological aftercare for all those who need it, and there's also time to discuss things in the canteen. There's a bowl of vegetable soup with mushroom-balls and a glass of nut milk ready and waiting for all of you."

7

It's the Law, Stupid!

If eating is a political act, then the law is a manifestation of belief. A deep belief in what it means to be human, and to be animal. About what pain is and how it feels to suffer. About what good behavior is, what is just, what is "right," and what isn't. The law is humanity's alternative Bible, the one that doesn't celebrate our belief in a supra-human god as a source of all goodness, but rather our belief in humanity as that source.

In this chapter I will show you that our laws, and our subsequent beliefs about humanity, have determined that we can feel free not only to use and eat animals but also to experiment on living animals, keep them locked up against their will in zoos and aquariums, and alter their appearance and genetics to fit our preferences. However, you will also see how the way we reflect on the difference between humans and animals has recently begun to radically change, and it is only a matter of time before this change becomes visible through law.

But don't get too excited: for a long time it wasn't knowledge but authority that determined which beliefs were turned into law, and which became punishable. The authorities of our laws are politicians, civil servants, philosophers, and scientists—a club of two-legged self-important "gods" often recognizable by their black suit coats and white lab coats. They make laws concerning people and other animals, and for centuries these laws have been supported by a popularly contentious belief: that humans are a subject—an "I" with a conscience—and animals an object—a thing.

Philosophical Debate

This philosophical idea was described in the Middle Ages by the theologian Thomas Aquinas, and later by influential thinkers such as Immanuel Kant in the eighteenth century. Aquinas believed that animals had no capacity for free will: he believed they were slaves to their environment, and as a result existed to be used as a kind of tool, not existing for themselves but for the benefit of others. According to Thomas, humans themselves did have free will thanks to their intelligence, thus making animals inferior to humans. Kant's view was remarkably similar to this. More specifically, he believed that only humans could be autonomous, because animals could have no sense of self-awareness or sense of rationality. Animals were therefore a resource, to be used by humans at will. This idea, that animals are tools for the use of humans, was dominant throughout the history of Christianity, and is deep-rooted in Western culture.

At the same time, this idea has always been contested. St. Francis of Assisi found it rather arrogant to believe that all life was put on Earth for man's benefit. He believed an animal's place on Earth was determined by divine plans, and not by human objectives, and so animals deserved to be treated equally. The eighteenth-century philosopher Rousseau argued that animals can feel and are very similar to us in many other ways, and so we should treat them well. We wouldn't want to treat other humans badly, so why would we do this to beings so similar to ourselves? After formulating his theory, Rousseau immediately decided to stop ordering around his dog. Retroactively he found the idea of "superior humans" acting as lords over animals very strange.

The best-known opponent of the belief that an animal is an object to be used by humans was the legal philosopher Jeremy Bentham (1748–1832). Bentham believed that it was not the capacity to reason logically but the ability to feel pain that should be the criterion by which we determine how we treat other animals. If the capacity for logical reasoning was the criterion for treating a living being well, then we would consider babies and certain mentally handicapped people as objects, as they cannot think for themselves either. Later thinkers on the topic of animal rights, such as Peter Singer and Tom Regan, built on this idea and argued that livestock farming, hunting, and animal testing should be stopped, as the suffering of animals did not outweigh the needs of humans.

Yet their arguments made little impression on our legal deities, and this age-old idea that animals are a tool is reflected in our laws as a result. Animals are thus not considered legal entities, meaning they have no legal rights or duties, but are rather an object, a thing.

Most people who begin thinking about this topic a little deeper start to find holes in that argument, as their intuition tells them there is something missing from the law. Surely a cow is different from a chair, right? If you tried to sit on a cow, she would either walk away or toss you off her back with a hefty buck: this is proof that, unlike a chair, she is alive, her heart beats, and there are certain things she likes and others she doesn't, as well as that she has her own free will. Yet according to the law, a cow is not a legal entity. She therefore cannot decide if someone has the right to sit on her back. This is her owner's decision. In this same way, she cannot decide if someone has the right to insert something into her vagina and impregnate her, or take her young calf away from her, or have her milk pumped out of her udders for commercial sale to humans. In this sense she is a tool; a possession that exists thanks to us, and this is the same for farm animals, lab animals, and pets.

My New Boyfriend's Balls

"Just get his balls cut off," my friend told me decisively as she arranged the lilies I had brought for her in a vase. "He's really stepped over the line, and it's only going to get worse." She stopped what she was doing and gave me a stern look. "Or if you don't want to do surgery, you can give him something that lowers his testosterone. It's no biggie. Before you know, he'll have calmed right down."

I had told her about the new love in my life, the one who had turned an adult woman into a giggling soft-hearted little girl when I had met him two years earlier during a work trip on a sunny Greek island, and whom I collected from the airport when I was back in the Netherlands two weeks later. During our first night in my apartment in Amsterdam neither of us caught a wink of sleep due to nerves, and the next day I was overjoyed as I showed him all my favorite spots around his new home. My favorite park, the beach where I liked to let off steam, my family, my friends, my life. From now on I would always have somebody when I got home from a long day of writing, or waiting for me when I arrived home jetlagged after a work trip.

But that's not how it turned out. A few weeks after his arrival, he ignored me for the first time: he didn't look up or around when I came back from an interview, he stood at the window with his back to me, or simply stared at other girls who walked past the house. In the days after that he started making noises at them too. Whining, squealing, whimpering . . . if a girl went past whom he found attractive, or who gave him even the slightest bit of attention, he stopped listening to what I was saying to him, even when I raised my voice. This carried on until twice, while we were out walking, he ran off and pounced on an unknown woman until I had to snatch him away from her. I asked some people around me for advice. They were all in agreement: now that my dog had reached puberty, he could no longer be trusted around bitches in heat, and so I should neuter him as soon as possible.

Their advice was born out of the widely accepted belief that if a pet doesn't behave according to how its owner wants, you have to make them adapt to what you want. Your pet is your property; your wish is his command. Often with the best intentions: dog owners, for example, want to prevent their horny adolescent dog from running out into the middle of a busy road out of sheer lust when he sees an irresistible dachshund bitch walking along the other side of the road. Better neutered than run over, right? And yet something seems illogical about our desire to make pets tamer than they are naturally: Why do we want dogs, who are the descendants of wild wolves, to live in our homes and then demand that they no longer behave like dogs? And not just that: we also want them to look cute, and not "wild."

We've already achieved that.

DIY Domestication

Charles Darwin wrote that all domesticated animals show remarkable similarities. They are all somewhat smaller than their ancestors, and their brains and teeth are also smaller. They often have drooping ears, curled tails, and white patches in their fur. They also look rather young, even long after they reach adulthood. The cutesy appearance of our present-day pets is the consequence of a centuries-long process of domestication. A number of wild animals—pigs, dogs, sheep, and rabbits—have been specifically bred and crossed by humans to make them more and more friendly. As a result, they have also become stupider and more dependent, as well as looking smaller, fluffier, and cuddlier.

In 2014, there were only around 200,000 wild wolves in the world, but more than 400 million domesticated dogs; there were 40,000 lions compared to 600 million house cats. In 2018, at least 60 percent of land mammals roaming the earth were farm animals, most of which were cows or pigs, which humans had bred for food or other uses. A mere 4 percent of land animals were "wild," that is to say, not domesticated by humans. Chickens and other poultry formed 70 percent of birds in the world: just 30 percent of birds are not bred, used, or eaten by us. Wild animals are increasingly in the minority, with an army of domesticated animals who are looking increasingly similar sticking it out.

It is also ridiculously easy to domesticate an animal. It's a much quicker process than you might think! This was proven by Professor Dmitry Belyaev, zoologist and geneticist, and his research assistant Lyudmila Trut, when they successfully turned an aggressive wild animal species into a tame and dependent pet within ten years. Their experiment concerned the silver fox, an animal so legendary for its aggressive behavior and penchant for biting that humans could only approach them while wearing special safety gloves.

Professor Belyaev preferred to delegate this task to his research assistant: every day, Lyudmila had to stick her gloved hand through the bars of the cage where a group of silver foxes was locked up and observe how they reacted to it. Beyond this, Lyudmila was not allowed to make any other contact with the foxes throughout the day: this way, the researchers would know for certain that any new behavior that emerged was the result of cross-breeding rather than from any learned tricks.

Lyudmila observed and made notes. She recorded which foxes reacted angrily to her hand and excluded them, and selected for further breeding only the foxes that reacted calmly to her hand.

Within four generations, the first fox began to wag its tail when it saw her. A couple of years later the foxes reacted to their names, and whined and barked for attention. They licked Lyudmila and her colleague's hand and wanted to play more and more, both with humans and with each other. The researchers knew that, in the wild, only baby silver foxes showed playful behavior: a month and a half after their birth, they became more serious and aggressive. The foxes in the experiment remained playful their whole lives. Not only did their behavior change but also their appearance. Their ears began to droop more with each generation, their noses got shorter, and so did their paws. The foxes developed curled tails, their fur became more colorful, male and female foxes began to resemble each other more closely, and females reached sexual maturity at an earlier age. They looked cute, and they began to behave more dependently, the result of being selected by the researchers because of a single characteristic: friendliness.

In 1999, a population of a hundred completely domesticated, dependent, well-behaved silver foxes had emerged. This silver fox 2.0 was used by humans because of its pelt, which was considered incredibly beautiful. In 2015, China was the world's biggest source of silver fox pelts, with 10 million being produced; in Europe that same year, according to the EFBA (European Fur Breeders' Association), this figure was 2 million pelts.

The domestication of dogs was a much slower process, but

the result was similar. In 2015, researchers from Oregon State University wanted to know what effect domestication had had on the behavior of dogs. To do this, they carried out a study that included wolves, domesticated street dogs, and household dogs. All three groups had an upside-down bucket placed in front of them, with a piece of meat placed under it. Eight of the ten wolves got to the meat without any problems. Most of the street dogs were unsuccessful, and nine out of ten of the household dogs didn't make any attempt to retrieve the meat and simply sat staring at their owners, who were present for the study.

Economic Losses

In recent years a lot has in fact changed about the way we think about animals, and as a result, things have also begun to change very slowly in our laws. Until a few years ago, cows and other animals literally appeared in the law as property, not as conscious beings with their own "I" that has worth and is distinct from their owner. They were "things" in the eyes of the law, in the same way that cars were: for protecting their owner from economic loss.

In the law, a car belonging to owner X cannot be used by person Y without the owner's permission: that is theft. The law also states that a car may not be damaged by someone who is not the owner: that is called third-party damage. If that car also happens to be a company car, and thus has value not only for the employee driving the car but also for the company, it may be stated in the law that the employee has a duty to look after their car so the company does not incur a loss. But the car itself has no rights and obligations—it is still an object.

Our laws concerning animals are based on the same principles: a cow has no rights, but someone who is not the owner of the cow is not allowed to do any harm to the cow, as this could lead to economic loss on the part of the farmer who does own it. Cattle farmers are also legally obliged to take care of matters concerning what is referred to in the law as the cow's "well-being," including, among other things, accommodation and transport. These laws were conceived in order to keep a cow as productive as possible within an economic system. In language, politics, and legislation, a cow is an object to be used by humans.

Since 2013, the law has changed slightly. The new changes under the new Animals Act in the Netherlands were more theoretical than practical. According to Article 1.3, animals are beings that can feel, and so any violation of the integrity and well-being of animals must be prevented, and according to Article 2.1, an animal may not be caused pain or harm and their health and well-being cannot be infringed. A change was also made to the Civil Code: it now states in Article 3.2a that animals are not objects. These seem like far-reaching changes. More specifically, this should mean that you cannot shoot a bullet through a cow's head, trim the beaks of chickens, burn the tails off piglets, gas healthy male chicks, insert needles into lab rats, allow elephants to do paintings in a circus tent, neuter dogs, keep killer whales in aquariums, or allow beginner riders to bounce up and down on the sensitive back of a horse.

But all of this continues to happen.

All legally.

It is allowed because the law does in fact state that you cannot cause an animal pain or infringe their well-being unless it serves a "reasonable purpose." This is allowed because, while in recent years Dutch and European law may have recognized that animals are capable of feeling, they are not treated as a subject and so are not given rights. And this is again allowed because while Article 3.2a of the Dutch Civil Code does declare that animals are not "objects," an extra line has been added that states, "Provisions relating to things are applicable to animals, with due observance of the limitations, obligations and legal principles based on statutory rules and rules of unwritten law, as well as of public order and public morality."

Excuse me?

What this means, as a legal researcher explained to me in everyday language, is "this amendment means that in our legal system, animals are still considered objects if that works out better for us humans."

There we go . . . no surprise that she and her colleagues saw barely any practical effect from making the "intrinsic worth" of animals legal. "There are some rulings on the books," she explained, "occasionally as part of divorce proceedings, or repossession. But this has hardly led to better protection for animals, and I see no structural improvements. Animals have no rights."

A cow cannot say if sitting on her back or artificially inseminating her serves a reasonable purpose, leaving it up to her owner to determine whether what he does with his cow is reasonable or not. And it is up to cattle farmers and consumers to determine whether how we deal with farm animals is reasonable or not. And it is up to scientists to determine whether it is reasonable or not for them to use an animal, and no other alternative, for testing medicines or other things.

It seems we find this reasonable millions of times a day.

Reasonable Legislation

According to Statistics Netherlands, there were 627,511,800 farm animals killed in slaughterhouses in the Netherlands in 2017. That is over 52 million per month, 13 million per week, 2 million per day, and, if we take a day to mean 24 hours, more than 78,000 per hour.

Reasonable.

In that same year, the Dutch Food and Consumer Goods Authority released a report that said every year 10 million chickens on Dutch poultry farms die before they are slaughtered. Another 1 million die during transport, 15 million suffer broken wings, they are not given enough water or food, and 1 in 5 have painful sores on their feet. The report also said that 99.9 percent of slaughterhouses regularly carry out slaughter carelessly: chickens are not always properly electrocuted, and as a result have their throats cut while they are still alive.

Reasonable.

In European zoos, around 3,000 to 5,000 healthy animals are killed each year because they are considered "surplus."

Reasonable.

In the twenty-first century, in tens of thousands of zoos and aquariums, animals are, as standard, locked up in spaces that are far too small for them: according to one well-known study, in an average zoo, lions and tigers have around 18,000 times less space than they do in the wild; for polar bears, this is 1 million times less. This means that animals in zoos, such as African elephants, killer whales, and lions, die much younger than those still in the wild.

Reasonable.

In 2016, 449,874 experiments were carried out using animal testing in the Netherlands alone. By the number of tests per animal species, this breaks down to 271,567 mice and rats, 52,237 chickens, 72,380 other birds, 28,476 fish, 10,129 pigs, 4,073 cows, 656 dogs, 438 sheep, 146 horses and donkeys, 70 rhesus macaques, 34 crab-eating macaques, 16 marmosets, 89 cats, 1,443 hamsters, 3,148 guinea pigs, 8,579 rabbits, 249 ferrets, and several hundred other animals.

In the United States that same year, there were 820,812 experiments using animals—and that's not even counting the 137,444 animals that were held in research facilities that year but were not used in any research, or the millions of fish, rats, and mice that were used, as in American law these have no right to protection of their "well-being" and are therefore not included in animal testing statistics. Altogether, the United States Department of Agriculture (USDA) and the Animal and Plant Health Inspection Service (APHI) estimate there to be a total of 12 million to 27 million vertebrates being used for testing in the United States.

Reasonable.

The footnotes of a report on the Dutch animal testing industry read, "It is possible for a test animal to be used for multiple studies. This means that the true number of test animals is lower than the number of animal tests carried out," and "According to the statistics, it appears that in 357,689 (88.7 percent) animal tests, the test animals are killed or die during the course of the study. In 45,681 (11.3 percent) tests, the animals were alive at the end of the study."

Reasonable?

Human Gods in White Coats

Humans rule over the law and the world. We lock up other animals, be it in cages, behind bars, behind barbed wire, in stalls, or in aquariums, because we think they exist for our benefit. We create more and more animals to satiate our demand for animal proteins; we modify them to our preferences and kill them when we feel they have served their purpose.

In 2018, various teams of scientists were in the process of creating pigs and chickens that grew faster from eating less food, while others attempted to create the perfect cow: a cow that would have the large udders of a dairy cow such as the industrious Holstein, but which could grow as fast as a meat cow. This super lucrative cow is yet to be born, as far as I could tell. These experiments, however, did lead to several hundred spontaneous abortions, miscarriages, and calves born with deformities. As I was writing this book, another research group was trying to develop pigs that, through modifying their DNA, would exhibit lower levels of stress in their blood. These levels had increased enormously in recent decades due to the increasingly "efficient" way we allowed farm animals to keep growing faster and faster, to transport them, and to slaughter them. These human gods in their white lab coats found the subsequent increase in the number of stress hormones in their blood caused by this rather irritating, not just for the pigs but also for the market (as stressed meat was tough meat, which left consumers dissatisfied).

You might think all that research money and so many top brains all in one place would have come up with the most obvious solution: if pigs and other animals get stressed by a certain type of livestock farming, transport, and slaughter system, something needs to be changed about that system. But that wasn't the solution these scientists put forward in their academic proposals; according to them, the thing that needed to change was the animals.

So what did scientists come up with as a solution to the problem of tough meat? They developed a DNA test that would help pig farmers prevent pigs who were too stressed (i.e., tough) from breeding any further. The test costs twenty euro and, according to a website where you can order the DNA test, a farmer is guaranteed a return on their investment, as stressed pigs can "cause particularly significant economic loss."

For tough beef, scientists came up with another, highly creative solution: so-called crate calves, removed from their mothers right after being born and kept in a crate only slightly larger than their own bodies. The calf then stays inside this crate until it is ready to be slaughtered, usually around four months later. It is not allowed to walk, play, or even move: the softer the muscles stay, the juicier and more tender the piece of veal on your plate will be.

The human gods deemed this to be reasonable.

Machines

What we believe is reasonable is linked to whether we believe animals feel less pain and suffer less than humans, as well as the idea that animals occupy a lower rung on the natural order ladder, i.e., the idea they are specially designed for us, as if we had originally conceived and created them . . . like Bitcoin or the steam engine.

Animals are also machines, wrote the French philosopher and mathematician René Descartes in the seventeenth century. They cannot think or feel: they are robots that react to what happens to their bodies with an automatic response. They are necessary devices, however: highly suitable for vivisection (cutting and opening up a living organism) so scientists can see what interesting things are going on inside their bodies.

When you start cutting into it with a knife, an animal will usually start flinching or screaming, but according to Descartes, this didn't mean the animal was scared or in pain; a tea kettle also makes a screaming noise when the water is boiling, but surely it doesn't feel pain.

His contemporary, Robert Hooke, one of the best-known scientists of the seventeenth century and a prominent member of the Royal Society, must have agreed with him. One day he wanted to know what the inside of a living being looked like when it was breathing, and to this end, one rainy afternoon, he tied a live dog to his examination table and cut the animal open: sawed through its ribs, pushed a hollow tube down its throat, and spent an hour examining it up close, fascinated by the up-and-down motion of its thorax, its wide-open eyes that kept staring at him, and its lungs, which inflated and then deflated. Hooke did not carry out his experiment with any pleasure: in a letter he later wrote to colleagues he mentioned how dreadful he found it to have made the dog suffer like that, but he saw it as a necessary evil, as well as his right. So it was, too. The law was on his side, and would stay that way for centuries to come.

Test Animals

Up until 1950, the Netherlands did not keep track of which animals were used in labs and what they were used for, or how they were treated. It was only in 1977 that the Experiments on Animals Act came into force. According to this law, you cannot carry out tests on animals without having a license for this that has been issued by an animal experiment commission. You can no longer dissect a living animal, but harming or killing them in other ways is still allowed, as long as members of the commission think the suffering of that animal is outweighed by the "need and necessity" of the experiment.

In the Netherlands, most animal testing is carried out to investigate the workings of the human body and physical and mental conditions. A smaller number are used for commercial research into medicine, or animal conditions, most of this research having the aim of ensuring farm animals do not suffer so much stress or illness that it makes their meat unsellable.

During an animal testing study, scientists have to make sure the animal test subjects contract the condition being studied, and then potential treatments are studied. For example, through genetic manipulation scientists ensure animals are born with a certain condition, or they make pregnant animals go through a treatment meaning they give birth to sick babies, or they inject a disease into healthy animals, or they use surgical intervention to compromise the organs or bones of these animals, or they create burns on their skins, or cause them to suffer depression, trauma, or anxiety disorders by applying electric shocks, or denying them food, water, sleep, or social contact.

Alternatives to Animal Testing

Not all researchers want to take part in this, however. In 2019 in the Netherlands, a small group of academics began developing alternatives to research using animal testing. Their labs are filled with trays containing living human tissue, and researchers are monitoring this very closely to see if it is developing the right way so it can be tested on later. Other researchers are embracing throwing themselves behind highly advanced computer technology that can be used as test material. The scientists are enthusiastic about the potential of their alternatives, and have high ambitions: they want the Netherlands to be "at the forefront of animal-free testing innovation."

The scientists who are currently working on this kind of research are doing so partly because they think animal testing is a miserable experience for the animals themselves. For example, during her neuroscience degree studies, PhD student Victoria de Leeuw had to replace the blood of living mice with paraffin, something she found so horrific she decided to dedicate the rest of her research to finding animal-free testing alternatives. Yet the most important reason is that a lot of animal testing doesn't actually appear to provide any relevant results for humans at all.

Animal tissues often react differently to substances applied to them than human tissues do: above all, animals often behave differently from how a human might in a similar situation. Countless studies have thus demonstrated tests on animals to be useless. Just think of the rats in the breast cancer study mentioned in chapter 6: soy appears to increase the risk of breast cancer among rats, but among humans it actually decreases it. This can be explained by the vast differences in hormone regulation between humans and rats.

These kinds of significant differences are being found more and more often. In 2011, researchers from the Radboud UMC Nijmegen, UMC Utrecht, and the Netherlands Heart Institute concluded that the vast majority of medical experiments on animals do not result in successful treatments for sick people. Up to 85 percent of all tests for new medicines and treatments had no effect on humans, while positive results had been observed on animal test subjects. A large proportion of these animal tests were also not carried out particularly well, with test subjects such as mice, rabbits, and primates dying unnecessarily. Other groups of scientists in the United States also came to the same conclusion.

Traditional

As I read through the dozens of licenses issued to researchers in recent years, I began to wonder who actually benefits from research using animal testing. For example, license application no. 20186086 wanted to conduct a study into schizophrenia in humans by first developing the condition among 3,744 mice and then using medication on these mice and conducting behavioral tests on them. For this experiment, the mice would be pricked with needles and other "negative stimuli," such as exposing animals that are naturally terrified of water to (deep) water. According to the researchers behind the application, this would lead to "moderate discomfort" among the mice, but would hopefully lead to "fundamental" insights into how schizophrenia works in humans. The application was approved and the study was carried out, with all the animals killed after the study.

Another application concerned a study into obesity among men and women in which the researchers wanted to expose 12,167 mice to surgical procedures, injections, and "treatments to cause an energy imbalance"; a vet explained to me this means the animals would be force-fed, or their environment would be made extremely cold, so that they would expend more energy but not get a chance to recoup it: they would effectively be starved. The application was approved and the study was carried out, with all the animals killed during or after the study.

Are these the kind of studies you would call "necessary"? Based on everything I had read on the lack of successful human results in animal testing, I highly doubt it. When the law suggests a study using animal testing can only be carried out if it is a necessary evil, but the "necessity" of this study seems, in many cases, to be questionable, where does that leave us? Surely only those animal testing studies that have been conclusively proven and defended as having saved or greatly improved human lives, and which absolutely cannot be carried out in any other way, right? Yes, say the developers of new animal testing alternatives. But they have spent years having to fight against a much larger group of fellow researchers who want to cling to their traditional working method: animal research.

For this reason it took until 2013 before the extraordinarily painful eye irritation test was scrapped (where potentially irritating or harmful substances were dropped straight into the eye of a live rabbit in order to observe what the effects would be), while we had known for centuries that animals could feel, and thus could suffer. Nowadays the eyes of dead chickens are used (waste from the meat industry). And what emerged from this? A chicken eye has a lot in common with a human eye, while a rabbit eye doesn't. The huge number of tests carried out on rabbits up until 2013 were therefore, in hindsight, rather unnecessary, as the results appeared to be completely irrelevant for humans. In an interview on her animal-free testing research, Victoria de Leeuw states that we will soon look back on all the tests being carried out on mice and other animal test subjects in much the same way: "Mice are also not entirely like us. In which case using models based on human cells would be much better."

I once spoke to someone who had to do testing on primates in a lab as part of her research. All the negative press about animal testing she had seen emerge in recent years bothered her. And to a certain extent I understood why. Even the staunchest opponents of animal testing I interviewed over the last few years agreed that, for the most part, lab animals have it much better than pigs, cows, and chickens that grow up in megabarns in order to be eaten by humans within a few months or years. Animal testing is, at least in theory, only allowed to be carried out if there are no animal-free alternatives, if you can demonstrate the need and necessity of the research, and if you submit an application. None of this is necessary for farm animals, which concerns a much larger group of animals, all while animal-free alternatives to dairy and meat have existed for a long, long time.

Unique Human Animals

You might have expected that we in the twenty-first century would have once and for all decided that Descartes was wrong about his age-old idea that an animal is a machine that can feel no pain, but that is not the case. There is still much discussion among scientists about whether animals can feel pain, and if so, how exactly, and if this is comparable to the way humans feel pain. Or, in the words of an information brochure about animal testing, "Pain is a difficult concept to attribute to animals, because we cannot ask them if they feel pain."

Well, no, you can't, of course, but if you ask me, my short answer would be go step on a cat's tail, see how it reacts, and then have a think about what you're asking. My long answer would be that, according to all reliable studies carried out on this subject, it appears an animal with a fully developed central nervous system has a sensory consciousness. This is at least the case for all vertebrates, i.e., mammals, fish, reptiles, amphibians, and birds. We know they experience pain in the same way humans experience it: their brains react the same, and so the scream your cat makes when you step on its tail cannot be compared to the sound of an old-fashioned kettle, but with the sound you make when you stub your pinky toe. Just like humans, animals react less intensely to pain stimuli when they have taken a painkiller, and more if the stimulus is unexpected. Even the smallest animals, such as crabs, snails, and fruit flies, are known to remember pain for a certain amount of time after experiencing it and to avoid it whenever they get the chance.

Research has also shown that animals not only feel physical pain but also experience a kind of mental suffering you can describe as a type of pain: depression, loneliness, extreme stress, and fear. From observations made on large-scale pig farms, we know that over time some pigs no longer respond to light or sound and no longer want to eat. They seem out of it and lethargic, behaving a little like humans locked in an isolation cell, or who have severe depression. Another study, this time into the well-being of animals in UK zoos, demonstrated that over half of elephants showed signs of serious stress. Zoo lions also spent almost half of their time each day pacing up and down along the bars of their enclosures: a kind of neurotic behavior that also points to stress.

If you take a goose's partner away from it, their stress hormones increase significantly. Giving them a new partner does not help, and the goose will only calm down once their partner comes back. Cows are extremely social animals and express strong emotions when it comes to relationships with other cows. It is known that in a herd, a cow makes two to four good friends and spends most of her time with those friends, showing this by licking their skin clean. Take these cows away, and the stress signals in their brains spike, they keep looking and calling for those cows, and they give off all kinds of other physical signals that, according to biologists, clearly show they miss their bovine friends. They also dislike other cows in the herd: if there's a dustup between two cows, for example, they can carry this resentment with them for months or even years and will want to stay out of each other's way.

As for killer whales, we know they spend their whole lives with their family or members of their pod. If children are removed from their mothers, in order to have to do tricks in an aquarium for an audience, for example, they experience similar physical and neurological signals that a human would experience when panicked.

Certain scientists counteract this by saying that animals do indeed feel stress and pain, but this pain is not as bad because animals cannot reflect on it like humans can. Well, I don't know how long you spend reflecting on pain when you stub your toe on something, but at that time I personally am usually too busy jumping up and down, screaming blue murder while clutching my injured toe. Other researchers seem to agree with me, as studies have shown that human pain is not a "reaction that demands much thinking power."

Humans and animals do not just have in common that they can experience pain and stress and will want to avoid those; they also share other deep-rooted needs, such as the need to not be locked up, the need to live in a group with other members of your species (in the case of herd animals such as humans, monkeys, dogs, pigs, rabbits, and chickens), the need to exhibit species-specific behavior (for example, pigs like to nibble on straw, while cows like to be able to move freely), and the need to bring up your own children and protect them from danger. We know that for both animals and humans, these ancient needs have no relation to high intelligence or self-awareness, but to something much deeper. This is why I feel scared when someone threatens to do me harm: an ancient instinct is telling me I need to protect myself somehow, and this is no different for a pig or a dog. This is also why a cow feels panicked and alone if you remove her suckling calf from her, because she is evolutionarily designed to live in groups and feed her offspring. You don't need any sort of IQ or human language for this. This is something you feel.

Humanopoly

Animals are similar to us not only in the way we experience pain and stress but also in most other areas. Although we have appointed ourselves lord and master of the Animal Kingdom, in reality we differ so little from animals that, according to scientists, there is no debate: humans are also animals, period. Except there *is* debate about this. A provocatively slow, dirtily played discussion that has been going on since the nineteenth century, when Charles Darwin proved with his theory of evolution that it is impossible to draw a hard line between humans and animals.

Since then, dozens of scientists have tried to win back humanity's monopoly. First, in the vein of Descartes, for a long time they said animals could not feel. After this they said they could not reflect on their feelings and behavior. Then they said they could not think about their own thoughts (metacognition in technical terms). Then that they were maybe capable of metacognition but not of moral thought. Then that they could not think about the future or the past. They had no free will, no language, no intelligence, and no deep emotions (such as regret, guilt, or shame). They couldn't use tools; they had no concept of time, no sense of self-consciousness, and they couldn't even recognize themselves.

Time and time again, all these ideas were convincingly disproved by animal researchers. Humans do occupy a special place, as biologists and developmental psychologists have shown time and time again, but one that is not above the Animal Kingdom, but somewhere in the middle of it. It is true we are an animal with special talents, but all animals have these. Humans are animals, but animals are not human. Every species of animal evolved in a way necessary for survival in their particular environment. Many animals are superior to us in terms of sight, smell, ability to swim, speed, endurance, power, tolerance of low and high temperatures, sociality, and other things they—and not us—have had to be very good at to be able to survive, and virtually all traits originally considered to be typically human are now known to also occur in the Animal Kingdom.

Self-recognition has been demonstrated in humans, great apes, dolphins, elephants, pigs, magpies and other corvids, and even ants. Human children develop this awareness around the age of two years. Dogs attach themselves to their owner, or to other beings. This, according to recent studies, demonstrates that dogs have a perception of themselves and their environment similar to that of a human child. Empathy, the ability to place yourself in another's shoes, also appears among all different kinds of mammals, such as monkeys, dolphins, and rats. This ability to empathize allows rats who don't know each other to put themselves in danger or forego rewards, even if this means suffering pain (fun fact: in behavioral studies on rats, the reward is often chocolate, as, just like humans, rats have a real sweet tooth). Their capacity for empathy was also demonstrated in studies where rats were placed next to other rats that had been locked in a cage, and learned both how to free these rats and how to find some way of getting a tasty chocolate snack. They would often free the other rat first and only then go for the chocolate. Half of the rats even shared it with the newly freed rat.

A slightly less fun fact about rats is that they are the most widely used animal in lab experiments. According to professor of drug metabolism and toxicology Geny Groothuis, one of the scientists trying to develop animal-free testing alternatives, that is because people do not find them cute and cuddly: dogs are not smarter or friendlier per se, but "they have a much higher cuddliness factor."

All kinds of animals use tools: monkeys, birds, fish, and octopuses.

Monkeys have a concept of time and can make plans for later in the day, or the day after. They wake up early if they know they have to compete for a good breakfast, and collect stones if they know they'll be useful for a certain task in a different location a few hours later.

Elephants, cetaceans, and ravens mourn the dead and conduct extensive mourning rituals.

A certain level of intelligence, sometimes surprisingly high intelligence, is evident among certain animals, even squid. The same goes for language. Of course, animals don't talk to each other in the same way as humans do, but they do indeed have their own language—or at least a language we can't understand. Gorillas can tell stories about their past using their own sign language; prairie dogs and squid seem to use a full-blown grammar, while bats like to gossip about one another. The fact they can't understand us, just like we can't understand them, doesn't automatically mean our "intelligences" are at different levels: they are simply different. When we hear someone from another country talking in their own language, we don't automatically decide they are stupider than we are, do we?

Although all illusions of humanity's superiority that have been fostered for centuries now seem to be falling apart, there are still a number of areas in which we outdo other animal species, such as creating and exterminating living animals.

Think about it: evolution, we laugh in your face! We humans are as skilled as God. He only created two humans and a garden full of animals, while we create billions of farm animals and pets.

We also score significantly higher when it comes to aggression. Well, yes, nature is cruel and wild animals fight one another, but only humans have developed weapons, technology, labs, and a range of other unfair methods to capture, torture, use, and kill other species of animals. Of all the wild mammal species that have ever lived on this planet over the last 4.5 billion years, only one-sixth remain. The rest have died out since the emergence of humans. A similar mass extinction is currently under way in our oceans: in 2017, only a fifth of the marine mammals remain of all the aquatic animals that existed until man decided everything in the sea belonged to him. Of all those sea and land mammals that have died out, half of them have gone extinct in the fifty years prior to my writing this book. Paul Falkowski, a highly acclaimed researcher from the US, called this a "unicum" of humanity: our capacity to systematically drive other species to extinction. Point to us, fellow humans.

Cecilia in Court

Steven Wise is an American lawyer whose organization, the Nonhuman Rights Project, tries to extend human rights to animals. Not because he believes animals are equal to humans in all aspects, or because he thinks that animals should have the same rights as humans, but because he thinks at the very least lawyers should stand for animals that are being badly treated by humans. This is not possible, however, as long as animals are viewed as a tool, and not as a legal entity.

Wise therefore tries to convince judges that certain animals in certain instances should be protected as a legal person, in the same way legal protection as a legal person can be given to a building or a company. It can even be given to a river or an area of forest: this has already been done in South America, New Zealand, and India as a way of protecting nature from humans who want to do her harm.

During all the years Wise has been doing this work, only one great ape, a chimpanzee named Cecilia, has obtained the status of a legal person. Cecilia lived in a concrete enclosure in a zoo in Argentina where she was very poorly cared for, and she now lives in a specialized sanctuary. According to Wise's website, she is doing well. Wise's colleagues have managed to obtain "nonhuman person" status for dolphins in India, and his team is now trying to do the same for an elephant in the United States.

In all the dozens of other lawsuits that Wise and his fellow animal rights lawyers have filed, their cases were thrown out, with some judges not even bothering to admit the case. Wise is still happy, however. "What we are doing is revolutionary," he told me in a phone call in the autumn of 2018. "We are at the point of radically changing the law. Judges still won't risk it, but we're now at a turning point. The lawsuit in which Cecilia obtained her status can now be used as a source of inspiration for other judges. This ape was the first animal in the United States to gain rights, but she won't be the last."

It seems, therefore, that Wise and other activists are having a tangible effect, one that often reaches far beyond the courtroom. In 2005, a zoo in Detroit became one of the first elephant-free zoos; in 2015, an Indian judge ordered a bird dealer to release the 500 or so birds he kept in cages, because "birds have a fundamental right to fly." In 2018, India's Supreme Court ruled that new poultry farms could no longer keep chickens or other birds in small cages; in 2019, two beluga whales from Changfeng Ocean World in China were flown to a sanctuary on the south coast of Iceland after the aquarium announced it would no longer be keeping dolphins and whales "because of ethical considerations."

In books and articles, Wise compares the current slow pace at which the law is changing regarding animals to the historical period in which laws around slavery changed. "Back then, people also said it wasn't possible," he explained. "They said our economic system would collapse and that people and companies would go bankrupt. The exact same arguments we're hearing now when it comes to animal rights. But it's inevitable that animals will ultimately get rights." A difficult task? It is, but for Wise it's simpler than that. "Animals cannot talk in a language that judges understand; if we don't speak for them, who will?"

Animals certainly cannot speak in human language. At least not yet.

Roanne van Voorst

Get Away from Me

In 2018, Amazon was in the process of developing a device that would allow us to read the thoughts of animals. They hoped to have such a device on the market within ten years: it's already been announced in the company's online catalog.

Place the device on your dog's head, and you may be able to hear it thinking something out loud like: "Oh yes, a gentle stroke on my head, yes please. Don't stop! Where'd the stroking go? What's that, food? She's going to the refrigerator! Oh dear, no food! Wait-wait-wait, look, let me show you a trick, give me some food!" Very amusing. Put this device on a dairy cow, however, and I imagine you'd hear her saying something like, "Get away from me. Don't touch that. Ow, that hurts! Go away. No! Where are you going with my child?" Not so amusing.

I cannot imagine farm or lab animals saying anything other than something like that if some way of translating their experiences into human language was discovered. What else do you expect? For a crate calf to secretly enjoy not having to bother with all that moving around, and regularly singing the farmer's praises? Or for pigs that get absurdly highly stressed during transport to all join in with singing songs in the back of the truck?

We have known for a long time now that we make decisions about animals that lead to them suffering—we just don't want to see it or talk about it or hear it. Not until technology forces us to listen, that is.

Given that billions of animals suffer pain and fear on a daily basis because we legally allow it, this invites the possibility of a difficult thought experiment: How can we live and eat in a way that does not exploit or wipe out other species? This thought experiment is difficult not only because it could mean we have to critically rethink our entire economic system but above all because it forces us to look at ourselves in a different way; to look at what a human is, and what a nonhuman animal is, and what animal rights could look like as a result, and how these differ from human rights, and the meaning of all these laws in relation to one another in an age in which humans have long since stopped behaving humanely.

Meet the Cyborgs

These are big questions that feel overwhelming, and subsequently a little terrifying. Yet not thinking about them at all and sticking to our current system of laws, in which humans have rights and other animals do not, is much more terrifying. If we state in our laws that we are allowed to rule over other beings because, according to these laws, we are superior to other animal species, then this means other beings can then rule over us if they believe we are below them.

These other beings are already among us. In 1998, British cybernetics professor Kevin Warwick became, in his own words, the world's first cyborg after he installed a radio chip in his arm. A cyborg is a physical compound of human and machine. Since his operation, he can turn on lights in his office just by snapping his fingers, doors open automatically for him because they can read his ID from a distance, and he can also manipulate a computer hand located on the other side of the world by controlling its fingers with his thoughts. For a time he even let his wife control his body with her brain (she has also been turned into a cyborg).

The couple is convinced that many more people must "upgrade" themselves, and quickly too, before we are overtaken by robots and computer systems developed by humans and that could become our rulers at any moment. Neil Harbisson, an artist living in New York, followed Warwick's advice. He wears an eyeborg, a color sensor, over his eye, which translates colors into sounds, allowing him to do something you and I cannot do: hear colors. When applying for a passport, he was allowed to wear the eyeborg on his head for the photo. Recognition of his cyber rights, or so Harbisson says.

What it means to be human will always keep changing. And the law will change with it. Today humans with a darker skin color are recognized as people, and in the past they were not. Today women are legally recognized as people with rights, but a couple of generations ago this was still not the case. In the Netherlands, someone born with a penis but who does not feel they are a man, but also not a woman, can now identify with a third gender, something not legally possible before 2016. Rivers and tropical forests have relatively recently been recognized as objects with intrinsic value, and have a right to be protected against anyone who wants to do them harm.

In the future, it seems that cyborgs will obtain human rights and animals nonhuman rights. What these rights will look like and how quickly they'll be implemented depends on the authorities we let determine our history at that moment in time. They will decide what research is needed and necessary, what suffering is considered "reasonable," and to what extent the suffering of another being outweighs our interests. Changes to the law now largely happen very slowly, but our behavior can change from one day to the next—as can the world. In the next chapter, you will see how global climate catastrophes can be avoided at the very last minute, as long as a very small number of people take action.

8

Melting Ice, Bursting Levees

Mud clung to my flip-flops. My toes were brown from the dirt and the trash. I leaned back against a wall, still warm from where the sun has been shining on it. If I craned my neck, I could see the top of the wall, about a foot and a half above me. Right behind me lay the Java Sea. The wall separated the water from the Indonesian slum where I was doing fieldwork. Every time I went there, I climbed on top of it; on one side, right next to my feet, was the sea. If I stretched my toes slightly over the stone, I could touch the water. On the other side, around six feet below, was the vast slum district. Muhammad the fisherman lived there with his two young sons and his darling wife, who always cooked water spinach for me when I came by because she knew I liked it a lot, in a house they had built themselves. And behind this? More houses, more slums, more children and more adults. To me the wall seemed comparatively thin for such a lofty task as protecting the people from sea flooding: I could stand on top of the wall with both feet together. "That thing is crumbling," Muhammad explained. "The last time the sea was rough, part of the levee broke and the water rushed into the house with such force it collapsed." The whole family survived, but Muhammad's wife broke her arm when she fell, and everything they owned was washed away.

Not somewhere you'd want to live, exactly. In 2014, water experts warned that Jakarta, Indonesia's capital city and a metropolis home to more than 10 million people, would be one of the first cities to disappear underwater as a result of climate change. I lived there for a year when I was doing research into the effects of climate change on the world's poorest people, where I found lodgings in one of the poorest and most flood-prone slum districts along the banks of the river in a house made from wood, mud, and corrugated iron. There was no running water: I had to get my water for cooking and washing from a well. Sometimes we had electricity, and even then only for a little while. When the river level was too high, all the generators stopped working, and suddenly everything was plunged into darkness.

It was a regular occurrence for me and other residents to have to flee onto the rooftops as extremely dirty river water streamed into our living rooms. The streets were flooded, and young boys paddled through the neighborhood in rubber dinghies to rescue the elderly and infirm from their homes. I personally found the times I experienced floods rather nerve-racking, but for long-term residents this was old news. The cheapest neighborhoods in the city were all nearest the river or the sea, and these areas were underwater numerous times during the year. This made the inhabitants sick, and even poorer every time the water swept away what meager belongings they had: the sofa they had saved up for, their children's school uniforms, the college funds squirreled away under a mattress; their business, their supplies of rice, their chickens . . . what was left behind on the floors of their self-built homes was a thick oozing residue, full of bacteria and other nasty things. No matter how well they scrubbed clean their houses and streets, after every flood people got sick or died and children suffered from strange symptoms: diarrhea, eczema, breathing difficulties, piercing headaches and nausea . . . these people had no money to go to the hospital, so people didn't know what the patient was suffering from, or even what they had died from. They had their suspicions, of course. Something to do with the many factories in Jakarta that dumped their chemicals in the city's rivers, or the fact that most people dumped their trash in public waterways, as the city had no well-functioning waste collection system in place.

On the coast of Jakarta, around an hour from where I lived while doing fieldwork on the future of climate changes, Muhammad lived with his family, as well as dozens of other fishermen's families whom I occasionally visited. Just like my neighbors, they too were plagued by floods, and the floods in this part of the city were also getting increasingly heavier. After every flood, the levee was repaired and raised by scrawny laborers who came and did the tough work for a pittance. Their bare torsos glistened with sweat, and gaps in their teeth were big enough to fit a cigarette through. They hauled broken bricks, poured them onto a layer of fresh cement, and bade the residents farewell, announcing, "You're all safe again for now, see you next time!"

They wouldn't have to wait long for that next time. During the period I was there, the people of Jakarta were threatened by a phenomenon the Dutch and millions of other people in low-lying areas of the world would soon also be threatened by. As seawater started warming in the last four decades, it began expanding. Sea levels are rising, much faster as a result of this, exceeding sea level rises due to the melting ice caps, and this in turn increases the global likelihood of flooding. Between 1900 and 2010, sea levels rose by more than 7 inches; the Intergovernmental Panel on Climate Change (IPCC) expects a further increase of between 17 and 29 inches by 2100. This is a conservative estimate. Famed climate scientist James Hansen and his fellow researchers predicted in 2015 that the sea will have risen "by several meters" by the end of the century. According to researchers at NASA, greenhouse gas emissions will be responsible for an increase in sea levels of at least 9 feet. Yet although scientists offered a solution that could prevent a climate crisis just in the nick of time, this message has not managed to get through to most consumers. For a long time it didn't get through to me either.

The Missed Message

Yeah, yeah, sure.

That's impossible.

That seems a little oversimplified to me.

It's totally overblown.

These were some of the thoughts that raced through my mind when I started reading a report on the impact of food on the climate a couple of years ago. The figures I read were so perplexing that during the first weeks of my research I honestly thought they were, to use the words of a certain US president, fake news.

I put the reports down.

I picked them up again.

I put them back down.

I e-mailed the scientists who had written them. "Dear sir/madam," I began. "Is this perhaps a misprint? An oversimplification so that the reader can understand the point better? Or has something gone way over my head and I just don't understand what you mean?" What had been written was no mistake, and my interpretation of the data was entirely correct. I then began to wonder if it was all a lie. Are these activists posing as "impartial scientists" and deliberately exaggerating in order to frighten readers or policy makers to push a particular political agenda?

It seemed not. If this were the case, there would have to be a whole lot of activists working for many different universities, and coming from all different parts of the world, working together toward some kind of secret agenda. All of their reports had similar messages. These messages could roughly be divided into four categories:

1. Climate change is natural and normal. This one isn't.

Climate change is a trending topic for us in the twenty-first century, but for Earth, it's business as usual. Colder and hotter periods are caused by changes in the amount of greenhouse gases in the atmosphere that absorb this heat, and changes in the scale at which the sun's rays are reflected back onto Earth's surface. These periods alternate every hundred thousand years or so, and used to be caused by natural circumstances: indeed, there were no other possible causes, as humans were not around for most of these climate change events. Yet while climate change is normal, the speed at which the climate has changed in recent years is not. Since the end of the last ice age (i.e., during the last 10,000 years), global temperatures have remained relatively stable, but since 1880 they have gone up by 0.13°F. The concentration of CO_2 in the atmosphere has increased by 40 percent between 1850 and 2018. Faster than ever.

2. Climate change is a fact; climate catastrophes are an option.

The increase in temperature depends on the increase of the amount of CO_2 in the air. According to IPCC scientists, there's a good chance that the warming of Earth can remain under two degrees if the amount of greenhouse gases in the atmosphere stabilizes to 450 parts of CO_2 per million parts of air. In January 2020, this was at 415, in which case sea levels would rise by more than a yard every 100 years, which would cause some issues in certain places but wouldn't result in a global catastrophe. The problem is that it would become impossible to remain under 450 parts unless we radically alter our behavior.

In 2050, it is estimated that Earth will have a population of around 9 billion people. If all these people eat and live in the same way that people in the West are currently used to, we're all officially screwed. Almost all scientists agree that such a scenario would lead to more extreme natural disasters, which would cause millions of climate refugees, injuries, and deaths, as well as increased levels of hunger, thirst, drought, famine, poverty, and armed conflict. The cost and devastation caused by this could be comparable to those experienced in the world wars and the economic depression of the first half of the twentieth century. Heat in cities will become unbearable and, depending on which models you consult, cities such as London, New York, and Amsterdam will end up underwater. According to researchers at the Global Challenges Foundation and the Future of Humanity Institute, there is a 5 percent chance extreme changes like this could be irreversible: they will set off a chain of events in motion that cannot be rectified and that will ultimately lead to humanity's demise.

3. A huge part of current climate changes are caused by the meat and dairy industry.

The reason behind this rapid increase in the amount of greenhouse gases is threefold: the Industrial Revolution, which began at the end of the eighteenth century; the invention of the combustion engine in 1867 and our large-scale burning of fossil fuels; and finally, the rise of factory farming. In the twenty-first century, the meat and dairy industries have had the most severe impact on the environment though the erosion of major grasslands due to overgrazing, increasing scarcity and pollution of water, loss of biodiversity, and deforestation as the result of the expansion of farmland. In 2019, agriculture occupied nearly 35 percent of the land on Earth, three-fourths of which was used for livestock farming, both for the livestock itself and for feeding it. Trees are cut down en masse to plant soy, maize, and other feed crops on the newly freed-up ground: in 2018, 16 million hectares of forest were destroyed (an area around twice the size of South Carolina). This deforestation released a lot of trapped CO_2. This output of greenhouse gases, caused by the livestock industry, was greater than from driving cars or flying, and also larger than the impact of the textile industry.

4. The most effective way of combating climate change is by turning vegan.

In 2017, a group of 1,500 scientists from 184 different countries called for people to eat vegan. Their message: a plant-based diet could save the world. If there is one thing individual people can do to prevent the worst possible disaster scenario, these scientists said, it is this: avoid eating meat and dairy as much as possible. Period. Other measures people could take that would have a major impact, such as buying fewer new things and flying less frequently, were ranked behind this.

The Good Old Days

There it was, in black and white. In report after report after report. And yet the idea that there is really only one thing we can do, and we have to do it *now*, still doesn't seem to be getting through. Not on January 1, when we are full of good intentions, not in five years' time when the kids have left home and we finally have time to experiment with new recipes, and not when vegan cheese tastes exactly the same as a nice bit of Old Amsterdam, which you enjoy with a lick of mustard and a glass of red wine, but *now*. Right now. Otherwise the world as we know it will soon cease to exist.

Otherwise you may very soon find a family of eight traumatized climate refugees sleeping on air mattresses in your house (there'll be so many, and they will have to go somewhere). Otherwise it will get so hot in the city where you live that, on top of worrying about all those climate refugees, you'll also have to go out at least three times a day with a jerrican to deliver water to your elderly neighbors, just to keep them alive. Otherwise it's more than likely you and your partner will fight more and more as a result of all this. Otherwise the number of ugly divorces will begin to increase sharply, just like the number of attacks against climate refugees, who will be increasingly—and openly—blamed for the economic crisis. Otherwise it is not unthinkable for a scenario to arise in which a politician emerges proposing that all these newcomers have to pay back all the health care funds they are claiming in this country through compulsory (and unpaid) labor toward building a sea wall. Then you will think back to the good old days, when the climate disaster was still an option, rather than a foregone conclusion.

The core of these scientists' messages did not get through to me for a long time, or to my friends, my family, most journalists, most of my fellow researchers, and also most politicians. Yes, there was something happening with climate change, something in far-flung parts of the world where we had always wanted to go on vacation some time, we knew that for certain. We read and heard about it so much that we would shut the paper with a sigh when we saw yet another photo of people in Bangladesh or Indonesia who had fled yet another flood. We made bad jokes about the potential of a typically damp, gray Northern European autumn being improved. "Climate change? Bring it on!" we exclaimed. "We could do with some warmer weather!" But we were far from realizing the food production system we had developed was the biggest driver of climate change, and that a plant-based diet could prevent such change from being so catastrophic.

The Deer-in-the-Headlights Effect

Now, don't you go thinking that people should start eating differently because "the end is nigh." News flash: the world will be perfectly fine. It's not Earth itself, but its inhabitants, that will die if we don't start living very differently very quickly. In the book *The World Without Us*, I read luscious descriptions of weeds climbing over the walls of abandoned houses, bursting through and filling any vacant nook and cranny—the writer observed this phenomenon in abandoned places all over the world, such as villages abandoned due to armed conflict. In these places, weeds dominated, houses tumbled down, animals made them home, cut-down forests grew back, and air quality improved. I devoured the first part of the book, but after that I grew bored. After 200 pages I got the message: in those abandoned places around the world that the book described, humans were not missed. *We* need nature to survive, but nature does not need us.

We simply cannot live without the oxygen that plants produce, and the earth in which we sow our crops. I think most people would therefore prefer the planet to stay healthy, because we need it to be for our own survival and for that of our children and our children's children. We know this rationally; we feel it too deep within us. Numerous studies have shown that our hearts beat more calmly when we are exposed to nature; we feel more relaxed, even by simply looking at a painting of a forest, a beach, or a beautiful flower. We should really try our best to treat Earth in a more sustainable way as a result, but the threatening messages about climate change we see in newspapers or on TV have a paralyzing effect on us. What we read is tough for us to take in. Too huge, too much, too complicated for us to sift out what the conclusion is. In other words: the light is too bright, it deprives us of sight, we can no longer see where we need to go, and so we close our eyes tight and we stand stock-still, waiting for what is coming toward us, like a terrified deer on a highway.

Scientific Disagreement

Something else that has ensured we ignore the advice of climate scientists is that the experts are still debating among themselves how many fewer animal products should be produced in order to prevent climate disasters. One school of thought is that we should stop eating meat and dairy altogether; the other is that we can still keep eating very small amounts of meat, butter, cheese, and eggs—and here we are waiting for the experts to finally reach a slightly clearer, more coherent conclusion, preferably in a language that people without specialized scientific titles can also understand.

Long story short: there isn't one.

Not because scientists disagree on the idea that the meat and dairy industry has an incomparable impact in the environment: that is an undeniable fact. The confusion stems from the emphasis placed on different aspects of the influence this industry has on the environment. Some look at emissions, while others find it more worthwhile to look at how to save the most land. According to one group of scientists, the latter would work best in a varied system of agriculture with a small number of farms that could take all the waste we do not use, consume it, and turn it into fuel or compost. More land would be freed up using a "reuse system," with small-scale livestock farming, than a completely plant-based agricultural system.

But other scientists posit that even a relatively small livestock farm can cause immense environmental problems. Ruminants will still be releasing methane, a greenhouse gas, no matter if they eat "waste plants" or feed grown especially for them, and so will make no difference in terms of emissions. Pigs and chickens have to be fed, transported, and slaughtered, all of which contributes to the release of greenhouse gases.

Then we still have the well-established facts that demonstrate animals are extraordinarily inefficient sources of protein. As inefficient as a broken adaptor in a plug socket.

The Broken Adaptor

In order to understand this metaphor, you first need to understand about proteins: they are chains of amino acids (it may help to imagine a string of beads). There are around twenty different amino acids, and the shape and structure of a protein is determined by which amino acids it contains and in what order they are "strung together." Humans and other animals cannot make all the amino acids on our own. We have to obtain some of our amino acids from our food. This could be from green plants, which can make their own amino acids, or from other plant-based protein sources such as grains or legumes. But this could also be from an animal that has first eaten these plant-based protein sources. This animal will convert these amino acids into proteins, which can then be extracted from their meat, milk, or eggs.

Farm animals therefore function as a sort of adaptor, like the ones you use on vacation in a country where the sockets have three holes instead of two. To get the current into your device, you insert a special adaptor into the plug socket. In principle, this is a well-thought-out system. In the case of food, however, this adaptor is not so good, particularly because animals do not convert all the food they eat into protein. They do, in fact, need a certain amount of them in order to grow, to breathe, and to move; only the surplus is left in their bodies.

Imagine I feed a chicken a large handful of grain, and I wait for her to lay an egg in order to make a delicious omelet. That period between feeding and laying, from grain to chicken to egg, involves a digestion process in which around 40 percent of proteins are lost. It would therefore have been much more efficient if I had just eaten the grain itself; then I would have absorbed all the proteins straight away. Chickens are relatively efficient protein adaptors compared to other farm animals, however; of the proteins that a cow consumes, 96 percent are lost in this same process.

We consume not only proteins in a relatively roundabout way but also fatty acids such as omega-3. We mostly do this by consuming sea life: for example, health officials recommend regular consumption of fish because they contain omega-3, fish oil in particular. It's certainly true that omega-3 is good for us. It has an anti-inflammatory effect, it is extremely important for a variety of brain functions, and it lowers the chances of getting certain forms of cancer, depression, arthritis, rheumatism, and dementia. Except fish do not make their own omega-3. They have it in their bodies because they eat seaweed, which does produce omega-3. We might have well just eaten that, rather than having to go through the fish first.

As we spent a long time wrongly believing that only animal products contained protein, as well as most people being convinced that we needed a lot more protein than we actually require (according to the WHO, on average, Western Europeans eat roughly twice as much protein as the average human needs; West Africans eat less protein than this, but still more than the average human needs), humanity has developed an extremely inefficient and, in fact, highly illogical food system. In 2018, meat and dairy products provided just 18 percent of our calories and 37 percent of the proteins we ate, while these industries used 83 percent of all agricultural land and were responsible for 60 percent of the greenhouse gases released by agriculture.

And That's the End of It

Land economist Joseph Poore of Oxford University's zoology department was fed up with all the scientific debating and arguing and wanted once and for all to get some clarity. Together with his Swiss colleague Thomas Nemecek, he carried out the most comprehensive study on food and climate change that had been done up to that point. In it he studied 38,700 farms all over the world, and more than 90 percent of all the food eaten by Earth's population at the time. Their conclusion, published in *Science* in 2018, stated that "avoiding animal products provides more advantages for the environment than everything else you could possibly do for the planet."

Poore found the results of his research so shocking that he immediately turned vegan and began to call for people to do the same. If people immediately stopped eating meat and dairy, global greenhouse gas output would be halved and water shortages and soil acidification could be reduced just enough to avoid the worst possible climate catastrophe.

According to Poore, simply becoming vegetarian, i.e., removing meat from your diet but continuing to eat eggs and dairy, is not enough: emissions from the cheese industry, for example, are comparable to those of the chicken and pork sectors. Becoming pescatarian, i.e., cutting out meat but continuing to eat fish, also doesn't do enough: the methane produced by fish farms is often higher than that of cattle, and 70 percent of the plastic found in the Pacific Ocean is waste from the fishing industry. A plant-based diet has more assured benefits for the environment than so-called sustainable dairy or meat products, because these are not environmentally friendly at all when compared to plant-based products. "Grass-fed beef" is often found on the menus of restaurants that want to present themselves as environmentally conscious, but in reality it is responsible for six times more greenhouse gas emissions than plant-based proteins (such as peas and beans) offering just as many nutrients, and uses thirty-six times more land. Poore calculated that even the most sustainably produced animal products are dozens of times more polluting than a plant-based product that generates relatively large emissions due to factors such as transport (as is the case for avocados or soy, for example).

His analysis held up and was convincing. Even the most ardent supporters of the reuse-land farming system admitted it, albeit grudgingly and only in the footnotes of their articles: a vegan diet offers far and away the most advantages for the environment, and even the most efficiently produced animals make somewhat less efficient use of the land than when we just plant edible plant-based proteins for human consumption straight into it. This argument, already settled in scientific circles, didn't make the news until the UN released a report that confirmed Poore's findings. All of a sudden, the papers ran headlines about veganism saving the world and urgent statements circulated on news sites declaring "a global transition to a vegan diet is crucial in order to save the world from hunger, energy shortages and the very worst impact of climate change." Or how about this: "the effects of agriculture will continue to increase because the global population is increasing and will be eating more animals . . . a substantial reduction in the effects on the environment is only possible through a substantial worldwide diet change, moving away from animal products."

Not Fair

It's downright unfair that the slum dwellers of Jakarta and those in other less developed countries will be the most threatened by the effects of climate change while, relatively speaking, they barely contribute to CO_2 emissions. The main cause of their problems is the relatively wealthy inhabitants of the planet, mostly living in areas that are still dry, comfortable, and safe—the inhabitants of this planet who use more energy than they do, who fly more often, and who have a much larger stake in the global meat and dairy industries.

In recent decades, for example, the Netherlands has imported a lot of livestock. Large industries first emerged in Europe and North America, and since the beginning of the Industrial Revolution, Europe's CO_2 emissions have been almost the same as Asia's, all while Europe has a population of 740 million compared to Asia's more than 4 billion. Europeans use far more than Asians do, and so it would be rightly fair for Europe (and the rest of the West) to take the lead and begin eating plant-based en masse because we have the opportunity to do so. This is a responsibility that comes with living in prosperous, methane-producing, greenhouse gas–emitting countries.

The reason why we have not immediately (and en masse) taken heed of the heartfelt pleas of scientists such as Poore is in part due to the fact that we prefer being lazy to being tired, and changing old habits requires a lot of effort. But it is even more in part due to our continually pointing the finger at other countries and other people: if they don't also start eating fewer animal products, surely it doesn't make sense for us to do it here, right? But that argument is not only weak; it doesn't make sense.

Don't get me wrong: a full global dietary transition is also completely unrealistic. If you live in a poverty-stricken or relatively isolated part of the world, you will have no access to affordable plant-based alternatives. Even in wealthy countries such as the United States you have so-called food deserts, where the poorest families in the country live and where plant-based products aren't readily available but where you can get a pound of chicken nuggets for a couple of dollars at a local fast-food drive-through. All these people aren't going to spontaneously switch over to seaweed sausages; they have other things on their mind. You also have rapidly growing economies, like China's, where plant-based food is very easy to get but where meat production has nonetheless rapidly increased in the last few years. The reason behind this is population increase, coupled with meat's status as a luxury product associated with the wealthy West. Until relatively recently, meat was completely unaffordable for the average Chinese person. By ordering it in a restaurant, you can demonstrate that you have made it in life. In 2014, Americans ate on average 264 pounds per person per year, the Dutch 198 pounds, and the Chinese 138 pounds. Some scientists therefore point out that the population of China and other countries where meat has gained this status in recent years will continue to increase, predicting that a plant-based future will have little chance of succeeding there.

Yet they forget that status is nothing more than fashion, and fashion is changeable. If an average Chinese person does indeed want to fill their belly with what "successful" Westerners eat, their preferences will change together with what food is fashionable here. This means that if hardcore meat-eaters are soon regarded as disapprovingly as groups of smokers currently are, plant-based food would also gain a higher status in China, and if eating oyster mushroom steak suddenly becomes trendy in the West, global demand for it will increase in turn. It will gain a "sexier" image and be associated with "good taste." If it is true that people in emerging economies will continue to eat meat because of its high status, it is up to us in the West to give plant-based foods a higher status.

Social norms in particular are contagious, as I made clear in chapter 3 when I discussed why eating plant-based had suddenly become cool. In that chapter, I also mentioned that scientific studies have shown that if 3 to 10 percent of a group of people can propagate a certain position with conviction, this will cause a ripple effect: more and more people will go along with it, and in the end a social shift will become unavoidable. This is because humans are social beings who do not like to be left behind. We prefer being part of a group of people who seem to know something for certain. This is the reason people adapt themselves to the status quo without having made a conscious decision to do so. If lots of people around you do something, you're also going to end up doing it. If you are born into a communist society, communism will feel normal to you and you will find it odd, bad, or even subversive to think differently. If everyone around you eats meat, it will feel like something normal and you will also end up eating animals. If you are born into a society in which eating animals is not an acceptable practice, however, you won't be so quick to break that taboo.

Maybe it would have helped if scientists and journalists had packaged it in a more positive way; instead, Poore, his fellow researchers, and the UN made plain to us in 2019 in no uncertain terms that we have *only* eleven years to fight the very worst effects of climate change. That was clearly the bad news. But the good news that came with this, and which completely went over our heads, was that we *still* have eleven years to combat the very worst effects of climate change, and if a relatively small group of people immediately stops consuming meat and dairy, *success* is achievable.

Epilogue

The Beginning of the End

During the writing of this book, my buying and cooking habits, my understanding of complicated questions about food and the environment, the way I look at my dog, the things I talk about with my partner during dinner, and the things I silently contemplate at lonely moments of the day all changed. The biggest change, however, was the questions that I asked, both of myself and of others.

During most of my research, I was looking for the answer to one question in particular: If, in theory, we could stop consuming animals, are we ready in practice to implement such a huge social, culinary, economic, and psychological change, or is that an unrealistic and idealist vision? Over the course of this research, I began to realize that this was not the right question. As humans, we are capable of doing anything if we really want to. History is full of examples of things we have achieved that were once considered inconceivable. We have:

- eliminated hunger from large parts of the world (with more people now suffering from being overweight than from hunger)
- abolished slavery and the burning of witches
- transformed homosexuality in the Netherlands from an "illness" and a crime into a sexual preference that is accepted by most people, all within thirty years
- taken twelve people to the moon, and developed two space probes that can travel so far they have now left our solar system (and we still have contact with them)
- conceived, drawn up, altered, and then eliminated borders between countries
- transformed livestock farming from farmers keeping a small number of animals to an industrial-scale factory farming industry worth billions of dollars

And so much more that seemed insane, unimaginable, utopian, or naive. The question I should be asking myself, therefore, is not *if* we can achieve huge social, economic, and psychological changes, but *which* change we as humans want to set in motion.

You should, too.

Two Extremes

Realistically speaking, we have a choice between two extremes. We can choose a world in which plant-based eating is the norm, or a world in which more and more animals are bred, imprisoned, and killed for food.

No change at all is not an option, although we often think it is. We tell ourselves it will all sort itself out; the air around us is still clean, the levees are still holding back the tide, we already choose organic meat at the supermarket most of the time, and there'll soon be a plant-based alternative to Greek yogurt that tastes exactly the same, and which will be so tasty that we'll start adding it to our shopping baskets more and more. That's the way we'd like to see the future: business as usual, but just a little more social, and a little greener.

But soon there will be no more business as usual. In the year 2100, in a worldwide high-tech economy of 9 billion human beings, it will be impossible to keep farm animals in humane conditions or in an environmentally friendly way. Land that is suitable for keeping animals is, for the most part, full—you just can't fit any more cows, sheep, or pigs on it. Feeding so many more people with animal proteins would require the development of vast factories, in which tens of thousands of animals will spend their lives crammed together in cages, and every second hundreds of thousands of pigs, chickens, cows, and other animals will be bred, used, and killed. In this scenario, smaller-scale meat and dairy businesses simply won't be able to survive in the face of large-scale, mechanized competition that is much more efficient and much cheaper.

After doing research into the future of the meat and dairy industries, Matthew Scully, a writer and former advisor to President George W. Bush, made a prediction that sent shivers down my spine: if we do not rebel against the way we treat animals in our society, the term "factory farming" will cease to exist within a couple of generations *because in a carnist future, no other way of keeping and slaughtering animals will exist.*

But it doesn't have to be this way. For centuries, eating animals was seen as morally acceptable because not only did we enjoy the taste, but we also needed them to survive—we had no plant-based alternatives available at the time. But this moral component has since disappeared. We no longer need animals to get proteins. We just think they taste good, and they're easy to use, and we've gotten used to it.

In order to choose a plant-based future, we don't need more money, time, or knowledge—there's plenty already.

The only thing we really need is to learn to see through the myths taught to us since we were children. Myths taught to us by doctors, our parents, our teachers, our food producers, and our politicians, i.e., people who also grew up with a carnist ideology, and who therefore believed that what they preached was normal, necessary, and natural. They believed that we needed animal proteins to survive and be healthy; that people have always eaten animals and will continue to eat them, and that animals feel no fear, pain, or stress when they are being slaughtered. They also believed perhaps the most dangerous myth of all: the idea we will never be able to run our economy and our kitchens any differently than we always have done.

We—the Rulers

The choice is up to you. Not to politicians, who are paid to do what their voters want them to do. Not to companies either, as they do what earns them the most money, and they do this by meeting the demands of consumers. Those consumers are you and me. We are the buyers, and we are the voters too. We therefore have the power to choose which of these two future scenarios becomes reality in the years to come.

There are of course political measures that can be created to move us toward that desired future more quickly. Things such as a meat tax, or forcing producers to list their emissions on food packaging and labels, or fines or other punishments for misleading labels such as "free-range" and "humane," or government subsidies for fruits, vegetables, and legumes, making plant-based living more affordable for everyone, or rewriting history books to give a more honest depiction of humankind's eating history, or modified advice from the organizations such as the Netherlands Nutrition Centre . . . but these measures have yet to be implemented, and as long as you and I don't show policy makers what we want through our actions and our behavior, they never will. Politicians and producers want voters and customers: they do what voters and customers want. Change must come from below. From us.

Many people find this a huge, terrifying responsibility; I find it a liberating and hopeful prospect. It reminds me I have the chance to assert my power, three times a day. Every time I eat is a vote for the future of my choice.

By simply continuing to buy products that are in line with the future I want to live in, I support companies that are in line with that future. I show supermarkets that I want to see more beans, nuts, oat milk, and Beyond Burgers on the shelves and that I don't need value packs of meat or cow milk. I show restaurants that I choose their plant-based options over their meat and fish ones. I show businesses that I'd rather spend my money at a vegan junk food spot than at a burger joint. They see this reflected in their sales figures. They report this to their investors, who then determine how and where they should invest their money over the coming years, and which start-ups they should support.

If you and I continue to pierce through the myths of carnism, recognize these marketing strategies as misleading, and live and eat in a way that makes our choice of future possible, we can make this happen in a relatively short period of time.

Temporary Discomfort

For many of us, a new way of thinking, eating, and living takes a lot of getting used to. We humans, however, have gone through worse things when we wanted to change the status quo because we no longer found it acceptable. People who helped escaped slaves risked huge fines and years in prison. Women who demanded the right to vote were ridiculed, had their ideas laughed at, and, if they got too aggressive, were arrested. During the writing of this book in our apartment in the US, thousands of Americans marched through the streets to protest discrimination and violence against people of color. They were cold, had had long days at work, and would have preferred to be watching Netflix than having to hold up signs with slogans and shouting at the top of their lungs. But they had to do it for their own sake; because they believed social change was more important than comfort. Around the same time, women from all over the world began sharing stories of sexual harassment and abuse under the hashtag #MeToo—stories that were shameful, and which were often met with painful criticism and skepticism from outsiders, but which managed to bring about shifts in gender and power imbalances.

And what do we risk? Cycling to a supermarket that is slightly farther away because your corner store doesn't have all the plant-based products you want. A batch (or three) of failed pancakes because you're still figuring out how to bind the mixture together without eggs. A child whining about what they used to eat, or a waiter frowning when you say you don't want today's special and ask for a vegan version of it. All of these are relatively uncomfortable, but nothing more than that.

But above all, this discomfort is *temporary*. The more people consciously choose to support one future and reject the other, the sooner that future will become reality. No magic required; just market forces. Supermarkets only want to stock their shelves with products that sell well. If consumers start filling their shopping carts with more and more plant-based products today, stores will begin stocking and promoting more and more of these products tomorrow. Food producers only want to supply products for which there is a demand (otherwise they incur a loss). All it takes is a few months of making an effort, and before you know it, our stores and supermarkets will be filled with new products from the future we wanted. We will get used to new recipes and foods, and our partners and children will too. Soon—sooner than you might imagine—the time will come when we can say: long long ago, we once ate animals.

Roanne van Voorst

The Future Starts Now

Whether it's climate change, the extinction of animal species, environmental pollution, or the ever-increasing suffering of ever more animals within the factory farming industry—all these are signs of us inching toward an abyss. They are signs that our economic model and the way we have fed ourselves for so long are now a threat to our existence and it is high time for a new way of living and thinking. They are signs that the stories we have told each other for so long no longer make sense and we must now create new stories that are better suited to our present situation.

As I am typing these words—and likely as you are reading them—we are both living in a vacuum between two stories. We know how things used to be, but we still can't quite imagine what things are going to be like when everything is different. In this book I have already sketched out most of this new story. All we have to do is start living it.

A study by the futurist Ian Pearson showed that it takes about thirty years for an idea that is initially labeled "impossible" to become reality. This figure was true for wide-reaching social changes from history that were ridiculed at first but achieved—and accepted—almost universally in the end. Today this process often happens much quicker: thanks to social media, trends can emerge within a day, and making the decision to go against a trend only takes a second. You can make that decision now. I could write many more chapters, ask myself and others many more questions, and then try to answer them in this book. But in the end, only one question really matters—one that you will no doubt be asked in the very near future, by your children, your nieces and nephews, your grandchildren . . . maybe even yourself.

When you knew at what point in history we stood, and what the consequences for the planet, animals, people in vulnerable areas of the world, and all future generations would be, what choice did you make? Whatever your answer is, I hope it begins, "Once upon a time, not so long ago, we ate animals."

It's time to evolve.

Acknowledgments

I've always found acknowledgments a little over the top. Authors often say they couldn't have written their books without Pete or Rose or whoever else, but that's baloney: they are the writer, and a writer writes with or without the input of Pete or Rose or whoever else. (Apologies to those of you I thanked in my previous acknowledgments; I thanked you for existing and for the healthy distractions you offered, but not for your writing help. You did not pick up a pen or go anywhere near a keyboard in connection with any of my books, so you do not deserve that honor. Don't get too annoyed about this, because the fact someone is happy you exist is also an enormous compliment.)

At least that's what I thought. I was wrong.

I wouldn't have been able to write this book at all without the help of a number of people. I could have attempted it, but I wouldn't have dared: I am far from an expert on these complex subjects, and I also didn't know if I even wanted to become an expert on this sensitive material.

At the top of the list of people who got me motivated to read and write the book is Lisette Kreischer. She proofread every word, pointed out embarrassing mistakes, sent me invaluable facts, congratulated me when I said I was making progress, and reassured me when at times I felt I was getting mildly obsessed or downright insane. This book would not have existed without her support, time, friendship, and knowledge.

Tobias Leenaert was another hero of mine over the course of writing this book. He stopped me from succumbing to "vegalomania" and made time in his already packed schedule to look over the manuscript. His comments helped me keep thinking critically about my own findings. I cannot thank him enough. Dietician Saraï Pannenkoek read the chapter about health and made sure I was more detailed where necessary, and completely truthful where possible. Fellow futurist Peter Joosten helped me look at the future even more clearly by reading through my fictional intermezzos.

I also want to thank my Dutch agent, Myrthe van Pelt, for carrying out complicated negotiations that I would never dare to do myself, for making sure my schedule was clear and calm, and for discovering that vegan champagne exists. Thank you to my American agent, Bonnie Nadell, for believing in this book. I want to thank Willemijn, Joost, and other amazing people at Uitgeverij Podium for their collaboration (and for the vegan cake), and HarperCollins for bringing this book to the rest of the world. Thank you to my fellow Freedom Writers Lou Niestadt and Roos Schlikker for their understanding of . . . well, everything, actually, and to Geertje Couwenbergh, who knew this was a writing adventure I needed to embark on alone.

Want to Know More?

In this book I have covered a lot of, but by no means all, topics surrounding a plant-based lifestyle. If you want to go deeper, I recommend the following sources.

Read

Jonathan Safran Foer, *Eating Animals* (Little Brown and Company, 2009). The most beautiful and honest book I have ever read about vegetarianism and veganism. For many people, it's the reason why they decided to go plant-based, and for me it was a huge source of inspiration when writing this book.

Jonathan Safran Foer, *We Are the Weather* (Hamish Hamilton, 2019). His follow-up book, about the relationship between what we eat and the climate. I interviewed Jonathan about this book for my podcast *The Braveheart Club* in 2019. You can find the episode, number 18, on Podbean or iTunes (all the other episodes are in Dutch, unfortunately).

Matthew Scully, *Dominion: The Power of Man, the Suffering*

of Animals and the Call to Mercy (St. Martin's Press, 2002). Another delightful and intelligently written book. Scully tackles not only factory farming but also commercial hunting and other activities where animals are mistreated. The interesting thing about this book is that Scully does not fit the stereotype of the left-wing, contrarian, anti-authoritarian activist: he is Republican, religious, and for many years was an advisor to George W. Bush. From this background he examines the way we treat animals, and concludes that this runs counter to everything that is (morally) right.

Tobias Leenaert, *How to Create a Vegan World: A Pragmatic Approach* (Lantern Books, 2017). This book is aimed at the vegan movement and animal rights activists. How can activists best communicate and take action in order to achieve their goal? By eating and using fewer animals? Leenaert's pragmatic approach and down-to-earth tone gave me a lot of inspiration during my own introduction to veganism. Leenaert has a less black-and-white vision than that of many other activists, and he is also a keen thinker who likes to challenge statements.

Melanie Joy, *Why We Love Dogs, Eat Pigs, and Wear Cows* (Conari Press, 2010). I featured Joy in this book in her role as a social psychologist, and if you want to read more about her ideas on carnism, I highly recommend her book. Her second book, *Beyond Beliefs* (Lantern Books, 2018), is very useful for mixed vegan/non-vegan couples and has lots of tips and information for nonviolent communication and how to influence effectively. Joy also gave a TEDx Talk (Toward Rational, Authentic Food Choices), which you can find on YouTube.

Peter Singer, *Animal Liberation* (HarperCollins, 1975; updated

2009). This is a standard reference work on veganism. Singer, a philosopher, gives a very logical explanation of why you should decide to stop eating animal products on ethical grounds.

If you want to read more about Black veganism, try Aph Ko and Syl Ko's *Aphro-ism: Essays on Pop Culture, Feminism, and Black Veganism from Two Sisters* (Lantern Books, 2017) and *Sistah Vegan: Black Female Vegans Speak on Food, Identity, Health and Society* (Breeze Harper, ed., Lantern Books, 2010). Also check out the Top 100 Black Vegans, AphKo: www.strivingwithsystems.com/2015/06/11/blackvegansrock-100-black-vegans-to-check-out.

Watch

I gave two TEDx Talks about the impact of the food industry on animal well-being and the climate, which also included tips on what you can do. You can watch the talks on my YouTube channel (bit.ly/roanneyoutube).

Blackfish (2013). This documentary deals with killer whales in captivity. The filmmakers interview former SeaWorld trainers and animal behavior experts, and reveal that the many claims made by commercial aquariums are not true, including the claim that wild killer whales live to be just as old as whales in captivity. This is not true: almost all whales in captivity die much earlier than those in the wild. Another claim that turns out to be incorrect is that killer whales "enjoy" learning and performing tricks for audiences. The film highlights a number of (deadly) incidents involving killer whales in captivity where they attacked their trainers.

The End of the Line (2009). This documentary, discussing the effects of fishing on the environment, animals, and people, is based on a book by environmental research journalist Charles Clover (who also appears in the film) and was lauded at a number of film festivals. It is available on Netflix.

Cowspiracy: The Sustainability Secret (2014). This documentary is not so much an examination of animal suffering in the meat and dairy industries as an investigation into the impact of the meat industry on the climate and the environment. Subjects discussed include water waste, CO_2 emissions, and land pollution and acidification. The film also examines how environmental organizations, but also the meat and dairy industries, deal with this information.

Earthlings (2005) and *Dominion* (2018). These two documentaries look at what happens to animals in the factory farming system. They are relevant if you want to see what happens on large-scale farms, but be warned: they include a lot of shocking images, often made by undercover activists inside livestock farms and slaughterhouses in the United States and Australia. Aside from the unpleasant images, one of the film's drawbacks is that it doesn't make clear how much of what you are watching are excesses or accidents, which left me with more questions than it did answers. While the film might not provide "the" truth (no creative work does), it does offer an insight into aspects of our food system that usually remain unseen. Watch, shudder, and judge for yourself.

The Last Pig (2017), a film about a pig farmer who decides he can't kill any more, and *73 Cows* (2018), a short documentary about Jay and Katja Wilde.

Cook

Deliciously Ella. This is my go-to for tasty, easy recipes. You can find most of them on Ella's website, www.deliciouslyella.com, or on her app.

Man.Eat.Plant. A blog by two Dutch women vegans, Lisette Kreischer and Maartje Borst, which will teach you hearty and nutritious plant-based recipes (www.maneatplant.com).

The men at *Wicked Healthy* offer more "manly" vegan recipes in their book and on their website (www.wickedhealthyfood.com); "manly" vegans should also check out Thug Kitchen (www.thugkitchen.com).

Vegan Challenge. Practical help for people who want to eat (more) plant-based but who are still a little unsure of how to get started (www.veganchallenge.nl).

If you are pregnant or have a family and want to find out more about veganism, Dreena Burton has great tips both on her website (www.dreenaburton.com) and in her book *Vive le Vegan!: Simple, Delectable Recipes for the Everyday Vegan Family*. I also recommend Alicia Silverstone and her blog *The Kind Life* (https://thekindlife.com/). She has also written a book on pregnancy and being a young mom, *The Kind Mama*.

Listen

Journalist Ezra Klein interviewed Melanie Joy for his podcast *The Ezra Klein Show*, becoming one of the clearest (and most recognizable) discussions about the shift to veganism I have ever heard. You can find this episode ("The Green Pill") on iTunes or wherever you get your podcasts.

Researcher and scientist Joseph Poore discussed the current situation of eating vegan and the climate on the *Deliciously Ella* podcast. The episode's title is "Veganism and Climate Change."

Philanthropist and former Citibank vice president Philip Wollen gave a speech on animal rights, which you can watch on YouTube under the not-so-modest title "Philip Wollen—Most Inspiring Speech on Animal Rights!"

Follow

The Joyful Vegan. American no-bull activist offering lots of free information and explanations on her Instagram feed.

Earthling Ed. An activist with polished points—a no-nonsense kind of guy. Watch the speech he gave to a room of university students, "You Will Never Look at Your Life in the Same Way Again."

Domz Thompson. The bag of muscles who "eats what elephants eat" provides a lot of food for thought (and muscles) on his social media platforms.

Moby. You can follow his (activist-esque) messages on Instagram, but I recommend checking out his TEDx Talk, "Why I'm a Vegan," on YouTube.

Do

Take part in the ProVeg Veggie Challenge! Your challenge is to spend thirty days eating healthier, tastier, more sustainably, and more cruelty-free. Whether that means eating vegan, vegetarian, or less animal proteins than you're used to is up to you and you only. You get a free coach, recipes, and practical tips, such as which supermarkets have good meat replacements on offer. Sign up at www.proveg.com/veggie-challenge/.

If you want to find out more information about my research work and writing, visit www.roannevanvoorst.com or find me on Instagram, @roannevanvoorst—I'd be happy to connect!

Notes

Introduction: Inventing a New Color

3 In 2018, Australia: See Murray-Ragg, Nadia, "Australia Is the 3rd Fastest Growing Vegan Market in the World." *Live Kindly*, January 23, 2018. See also "Vegan Trend Takes Hold in Australia." *SBS*, April 1, 2018. See also Wan, Lester, "Fact Not Fad: Why the Vegan Market Is Going from Strength-to-Strength in Australia." *Food Navigator Asia*, April 25, 2018. See also: "Top Meat Consuming Countries in the World." *World Atlas*, April 25, 2017.

3 In the United States, not only have: This appeared in research carried about by Nielsen for the Plant Based Foods Association (PBFA) and the Good Food Institute.

3 By 2021, these alternatives: Fox, K. "Here's Why You Should Turn Your Business Vegan in 2018." *Forbes*, December 27, 2017. See also Packaged Facts.

4 Dairy Farmers of America: See "Dairy Farmers of American Reports 1 Billion in Losses in 2018." *The Bullvine*, March 22, 2019.

4 In January 2019, the Canadian Food Inspection Agency published: Kirkey, Sharon, "Got Milk? Not so Much. Health Canada's New Food Guide Drops 'Milk and Alternatives' and Favours Plant-Based Protein." *Canada News Media*, January 22, 2018.

4 By 2024, its estimated global: This is based on a report by market research provider Bharat Book.

5 The Dutch newspaper *De Volkskrant* considered: Eerenbeemt, Marc van den, "De opmars van de vleesvervangers zet door: Unilever koopt Vegetarische Slager." *De Volkskrant*, December 19, 2018.

6 In 2017, 42 percent: This was a calculation by the Vegan Society, the oldest vegan association in the world.

10 The idea that: See Hedges, Chris, "What Every Person Should Know About War." *New York Times*, July 6, 2003. See also Barwick, Emily Moran, "How Many Animals Do We Kill Every Year?" *Bite Size Vegan*, May 27, 2015.

12 Historian Yuval Noah Harari wrote: Harari, Yuval Noah, *Sapiens: A Brief History of Humankind*. New York, HarperCollins 2015.

15 our apple juice: "Verrassing: in appelsap zit vaakvarkensvlees." *Joop*, October 2, 2016.

Chapter 2: Why Good People Believe in Bad Stories

43 found in Ethiopia: Hublin, Jean-Jacquesen Abdelouahed Ben-Ncer, Shara E. Bailey, Sarah E. Freidline, Simon Neubauer, Matthew M. Skinner, Inga Bergmann, Adeline Le Cabec, Stefano Benazzi, Katerina Harvati, and Philipp Gunz, "New Fossils from Jebel Irhoud, Morocco and the Pan-African Origin of *Homo Sapiens*." *Nature* 546 (2017): 289–292.

52 A large-scale study from 2017: Reese, Jacy, "Survey of US Attitudes Towards Animal Farming and Animal-Free Food October 2017." *Sentience Institute*, November 20, 2017.

54 and increasing affluence: "Zuivelindustrie." *ABN AMRO Insights*, April 13, 2017.

63 They were "different": "Tijdbalk Vrouwenkiesrecht." *Vereniging voor Gendergeschiedenis*.

Chapter 3: From Pasty and Peeved to Sexy as Fuck

82 *Alas, what wickedness*: Walters, Kerry S., and Lisa Portmess, eds., *Ethical Vegetarianism: From Pythagoras to Peter Singer*. Albany, NY: State New York Univ. Press (1999), 13–22.

Chapter 4: Giraffes for the Rich, Vegetables for the Poor, and Milk for All

104 A cookbook from the Middle Ages: Scholier, Peter, *Koock-boeck ofte familieren kevken-boeck*, 1663.

106 The Dutch Milk Board: See Zuivel online.

108 according to a 2014 sector report: See Zuivel online.

114 who made sure kale: Meijer, Anke, "De groteboerenkoolhype." *NRC Handelsblad*, July 12, 2014.

115 and others did the same with: Thole, Herwin, "Superfoods zijn pure marketing, daar prikt de Keuringsdienst genadeloos doorheen." *Business Insider*, April 30, 2015.

118 Heme molecules are: Brodwin, Erin, "Silicon Valley's Favorite Veggie Burger Is About to Hit a Wave of Controversy—but Scientists Say It's Bogus." *Business Insider*, April 20, 2018. See also: Hincks, Joseph, "Meet the Founder of

Impossible Foods, Whose Meat-Free Burgers Could Transform the Way We Eat." *Time*, April 23, 2018.

118 Chang, a huge fan of: "David Chang on Veganism and the Environment." Big Think, April 23, 2012.

Chapter 5: Wanted: Man (20–40), Sporty, Sexy, Vegan

127 in the year before: "Hoeveel dieren zijn er in 2017 geslacht in Nederland?" *VATD Blog*, June 26, 2018.

140 "fire, glowing coals": Vugts, Pascal, "Waarom mannen barbecueën." *Hoe mannen denken*, May 28, 2016.

Chapter 6: Plant Overdose

148 No surprise, therefore, that one in three: "Vandaag: protest tegen de dieetindustrie in Engeland." *Wondervol*, January 16, 2012.

148 They fully understand: Markey, Charlotte, "5 Lies from the Diet Industry." *Psychology Today*, January 21, 2015.

149 So much so that: Jonkers, Aliëtte, "Hoe gezond zijn vleesvervangers?" *De Volkskrant*, December 18, 2016.

153 regular consumption of which: Li, Fei, Shengli An, Lina Hou, Pengliang Chen, Chengyong Lei, and Wanlong Tan, "Red and Processed Meat Intake and Risk of Bladder Cancer: a Meta-Analysis." *International Journal of Clinical and Experimental Medicine* 7, no. 8 (2014): 2100–2110. See also Steck, Susan, Mia Gaudet, Sybil Eng, Julie Britton, Susan Teitelbaum, Alfred Neugut, Regina Santella, and Marilie Gammon, "Cooked Meat and Risk of Breast Cancer—Lifetime versus Recent Dietary Intake." *Epidemiology* 18, no. 3 (2007): 373–382. See also Rohrmann, Sabine, et al., "Meat Consumption and Mortality—Results from the European Prospective Investigation into Cancer and Nutrition." *BMC Medicine* 11, no. 63 (2013). See also Aune, Dagfinn, Doris S. M. Chan, Ana Rita Vieira, Deborah A. Navarro Rosenblatt, Rui Vieira, Darren C. Greenwood, Ellen Kampman, and Teresa Norat, "Red and Processed Meat Intake and Risk of Colorectal Adenomas: a Systematic Review and Meta-Analysis of Epidemiological Studies." *Cancer Causes & Control* 24, no. 4 (2013): 611–627. See also Zwaan, Juglen, "6 wetenschappelijk onderbouwde voordelen van veganistisch eten." *A Healthy Life*, February 9, 2017.

153 research has indeed shown: Oyebode, Oyinlola, Vanessa Gordon-Dseagu, Alice Walker, and Jennifer S. Mindell, "Fruit and Vegetable Consumption and All-Cause, Cancer and CVD Mortality: Analysis of Health Survey for England Data." *Journal of Epidemiology and Community Health* 68, no. 9 (2014): 856–862. See also Herr, I., and M. W. Büchler, "Dietary Constituents of Broccoli and Other Cruciferous Vegetables: Implications for Prevention and Therapy of Cancer." *Cancer Treatment Reviews* 36, no. 5 (2010): 377–383. See also Royston, K. J., and T. O. Tollefsbol, "The Epigenetic Impact of Cruciferous Vegetables on Cancer Prevention." *Current Pharmacology Reports* 1, no. 1 (2014): 46–51.

153 A third reason that ensures: Zhang, Caixia, Suzanne C. Ho, Fangyu Lin, Shouzhen Cheng, Jianhua Fu, and Yuming Chen, "Soy Product and Isoflavone Intake and Breast Cancer Risk Defined by Hormone Receptor Status." *Cancer Science* 101 (2001): 501–507. See also Wu, Anna H., Peggy Wan, Jean Hankin, Chiu-Chen Tseng, Mimi C. Yu, and Malcolm C. Pike, "Adolescent and Adult Soy Intake and Risk of Breast Cancer in Asian Americans." *Carcinogenesis* 23, no. 9 (2002): 1491–1496. See also "Eten en kanker: de broodnodige nuance." *Gezondheidsnet*, December 8, 2016. See also Katan, Martijn B., *Voedingsmythes: over valse hoop en nodelozevrees*. Amsterdam: Prometheus/Bert Bakker, 2016.

154 These studies also showed: *Soja en borstkanker: wat ishet verband?* Wereld Kanker Onderzoek Fonds.

154 studies that compared: Le, Lap Tai, and Joan Sabaté, "Beyond Meatless, the Health Effects of Vegan Diets: Findings from the Adventist Cohorts." *Nutrients* 6, no. 6 (2014): 2131–2147. See also Oyebode, Oyinlola, Vanessa Gordon-Dseagu, Alice Walker, and Jennifer S. Mindell, "Fruit and Vegetable Consumption and All-Cause, Cancer and CVD Mortality: Analysis of Health Survey for England Data." *Journal of Epidemiology and Community Health* 68, no. 9 (2014): 856–862. See also Casiglia, Edoardo, et al., "High Dietary Fiber Intake Prevents Stroke at a Population Level." *Clinical Nutrition* 32, no. 5 (2013): 811–818. See also Threapleton, Diane E., Darren C. Greenwood, Charlotte E. L. Evans, Cristine L. Cleghorn, Camilla Nykjaer, Charlotte Woodhead, Janet E. Cade, Chris P. Gale, and Victoria J. Burley, "Dietary Fiber Intake and Risk of First Stroke—A Systematic Review and Meta-Analysis." *Stroke* 44, no. 5 (2013): 1360–1368. See also Bazzano, Lydia A., Jiang He, Lorraine G. Ogden, et al., "Legume Consumption and Risk of Coronary Heart Disease in US Men and Women: NHANES I Epidemiologic Follow-up Study." *Archives of Internal Medicine* 161, no. 21 (2001): 2573–2578. See also Nagura, Junko, Hiroyasu Iso, Yoshiyuki Watanabe, Koutatsu Maruyama, et al. "Fruit, Vegetable and Bean Intake and Mortality from Cardiovascular Disease among Japanese Men and Women: the JACC Study." *British Journal of Nutrition* 102, no. 2 (2009): 285–292.

154 Various studies have also: Dinu, Monica, Rosanna Abbate, Gian Franco Gensini, Alessandro Casini, and Francesco Sofi, "Vegetarian, Vegan Diets and Multiple Health Outcomes: A Systematic Review with Meta-Analysis of Observational Studies." *Critical Reviews in Food Science and Nutrition* 57, no. 17 (2017): 3640–3649. See also Mishra, S., J. Xu, U. Agarwal, J. Gonzales, S. Levin, and N. D. Barnard, "A Multicenter Randomized Controlled Trial of a Plant-Based Nutrition Program to Reduce Body Weight and Cardiovascular Risk in the Corporate Setting: The GEICO Study." *European Journal of Clinical Nutrition* 67, no. 7 (2013): 718–724. See also Macknin, Michael, Tammie Kong, Adam Weier, Sarah Worley, Anne S. Tang, Naim Alkhouri, and Mladen Golubic, "Plant-Based No Added Fat or American Heart Association Diets, Impact on Cardiovascular Risk in Obese Hypercholesterolemic Children and Their Parents." *Journal of Pediatrics* 166, no. 4 (2015); 953–959. See also Wang, Fenglei, Jusheng Zheng, Bo Yang, Jiajing Jiang, Yuanqing Fu, and Duo Li, "Effects of Vegetarian Diets on Blood Lipids: A Systematic Review and Meta Analysis of Randomized Controlled Trials." *Journal of the American Heart Association* 4, no. 10 (2015).

154 This is particularly good news: Lu, Y., K. Hajifathalian, et al., "Metabolic Mediators of the Effects of Body-Mass Index, Overweight, and Obesity on Coronary Heart Disease and Stroke: A Pooled Analysis of 97 Prospective Cohorts with 1.8 Million Participants." *Lancet* 383, no. 9921 (2014): 970–983. See also Tonstad, Serena, Terry Butler, Ru Yan, and Gary E. Fraser, "Type of Vegetarian Diet, Body Weight, and Prevalence of Type 2 Diabetes." *Diabetes Care* 32, no. 5 (2009): 791–796. See also Gojda, J., J. Patková, M. Jaek, J. Potoková, J. Trnka, P. Kraml, and M. Andl, "Higher Insulin Sensitivity in Vegans Is Not Associated with Higher Mitochondrial Density." *European Journal of Clinical Nutrition* 67 (2013): 1310–1315. See also Le, Lap Tai, and Joan Sabaté, "Beyond Meatless, the Health Effects of Vegan Diets: Findings from the Adventist Cohorts." *Nutrients* 6, no. 6 (2014): 2131–2147. See also Craig, Winston J., "Health Effects of Vegan Diets." *The American Journal of Clinical Nutrition* 89, no. 5 (2009): 1627S–1633S. See also Barnard, Neal D., Joshua Cohen, David J. A. Jenkins, Gabrielle Turner-McGrievy, Lise Gloede, Brent Jaster, Kim Seidl, Amber A. Green, and Stanley Talpers, "A Low-Fat Vegan Diet Improves Glycemic Control and Cardiovascular Risk Factors in a Randomized Clinical Trial in Individuals with Type 2 Diabetes." *Diabetes Care* 29, no. 8 (2006): 1777–1783. See also Morita, E., and S. Fukuda, J. Nagano, et al., "Psychological Effects of Forest Environments on Healthy Adults: Shinrin-Yoku (Forest-Air Bathing, Walking) as a Possible Method of Stress Reduction." *Public Health* 121 (2007): 54–63. See also Pearson, D. G., and T. Craig, "The Great Outdoors? Exploring the Mental Health Benefits of Natural Environments." *Frontiers in Psychology* 5 (2014): 1178. See also Mackay, J., and G. N. James, "The Effect of 'Green Exercise' on State Anxiety and the Role of Exercise Duration, Intensity, and Greenness: A Quasi-Experimental Study." *Psychology of Sport and Exercise* 11 (2010): 238–245. See also Coon, J. Thompson, K. Boddy, K. Stein, et al., "Does Participating in Physical Activity in Outdoor Natural Environments Have a Greater Effect on Physical and Mental Wellbeing than Physical Activity Indoors? A Systematic Review." *Environmental Science & Technology* 45, no. 5 (2011): 1761–1772.

157 While traditionally produced tempeh: "Meer vitamine B12 in lupine tempé door in-situ verrijking." Wageningen University and Research, January 1, 2016—December 31, 2018.

161 *in developing countries where*: Zwaan, Juglen, "8 signalen en symptomen van een eiwittekort." *A Healthy Life*, June 14, 2018. See also Hamilton, Lee, "Welke eiwitten zijn het best voor spieropbouw—dierlijke of plantaardige?" *EOS Wetenschap*, April 5, 2017.

161 from animal or plant sources: Mangano, Kelsey M., Shivani Sahni, Douglas P. Kiel, Katherine L. Tucker, Alyssa B. Dufour, and Marian T. Hannan, "Dietary Protein Is Associated with Musculoskeletal Health Independently of Dietary Pattern: The Framingham Third Generation Study." *The American Journal of Clinical Nutrition* 105, no. 3 (2017): 714–722.

163 Much later, *Homo erectus*: Melamed, Yoelen, Mordechai E. Kislev, Eli Geffen, Simcha LevYadun, and Naama Goren-Inbar, "The Plant Component of an Acheulian diet at Gesher Benot Ya'aqov, Israel." *PNAS* 113, no. 51 (2016): 14674–14679.

167 A healthy baby: Berkel, Rob van, "Is melk een probleem door lactose-intolerantie?" *Over voeding en gezondheid*, November 19, 2014.

168 Professor Caleb Finch: Finch, C. E., and C. B. Stanford, "Meat-Adaptive Genes and the Evolution of Slower Aging in Humans." *Quarterly Review of Biology* 79, no. 1 (2004): 3–50.

Intermezzo: A School Trip to the Slaughterhouse

172 This was when they were: "Slachtdoordacht–optimaal slachtgewicht." *Varkensloket*.

172 "Yes," the guide says, "it had been": "Misstand #74: Afbranden/knippen van biggenstaartjes." *Varkens in Nood*.

173 "and because pigs were": "Boeren omzeilen verbod op afbranden varkensstaartjes." *Metro*, November 1, 2016.

173 In those early years after: See: "Veelgestelde Vragen." *Hobbyvarkenvereniging*, 2019.

175 "In the Netherlands, the egg industry": Heck, Wilmer, "Nederland hakt of vergast 30 miljoen jonge haantjes per jaar." *NRC*, May 7, 2012.

175 "As for the pigs": "Zo doden Nederlandse slachters jaarlijks miljoenen varkens." *NOS*, March 28, 2017.

175 "In 2013, the average weight of a pig": "Meer varkens en 1,5kilo zwaarder geslacht in 2017." *Varkens.nl*, January 4, 2018.

178 Products with a dairy-related name: Keuken, Teun Van De, "Het grote 'verwarringsgevaar': Sojamelk mag geen soja melk meer heten." *De Volkskrant*, July 3, 2017.

179 soon after the company: Dinerstein, Chuck, "Is a McCricket the Breakfast of Our Future?" *American Council on Science and Health*, August 13, 2018. See also "McDonald's komt met McVegan: 'Over 15 jaar zijn alle snacks vega'" *NOS*, December 19, 2017.

Chapter 7: It's the Law, Stupid!

187 Chickens and other poultry: Carrington, Damian, "Humans Just 0.01% of All Life but Have Destroyed 83% of Wild Mammals—Study." *The Guardian*, May 21, 2018.

187 This was proven by: Bregman, Rutger, "Hoe de mens de baas op aarde werd." *De Correspondent*, August 4, 2018.

188 according to the EFBA: see the Fur Europe website.

189 To do this, they carried out: Udell, Monique A. R., "When Dogs Look Back: Inhibition of Independent Problem Solving Behaviour in Domestic Dogs (*Canis lupus familiaris*) Compared with Wolves (*Canis lupus*)." *Biology Letters* 11 (2015).

191 What this means: Graaff, R. L. de, "Dieren zijn geenzaken." *Ars Aequi*, September 2017.

193 By the number of tests per animal species: "USDA Publishes 2016 Animal Research Statistics." *Speaking of Research*, June 19, 2017.

193 In the United States: "USDA Publishes 2016 Animal Research Statistics—7% Rise in Animal Use." *Speaking of Research*, June 19, 2017.

194 "According to the statistics": "Vleesproductie; aantal slachtingen en geslacht gewicht per diersoort." *CBS*, January 31, 2019.

194 through modifying their DNA: See Scully, Matthew, *Dominion*, p. 236. See also the VHL Genetics website.

197 The law was: Brief van Robert Hooke aan Robert Boyle (November10, 1664). In Hunter, M., A. Clericuzio, and L. M. Principe (ed.), *The Correspondence of Robert Boyle* (2001), vol. 2, 399.

199 Up to 85 percent of all tests: "De Peiling: Honderdduizenden proefdieren sterven nutteloos." *NH Nieuws*, July 5, 2018.

210 It can even be given to a river: Safi, Michael, "Ganges and Yamuna Rivers Granted Same Legal Rights as Human Beings." *The Guardian*, March 21, 2017.

210 Wise's colleagues: "Four Reasons Why India Recognises Dolphins as 'Non-P. 190' In 2005 werdeen . . .' Schweig, Sarah V., "Smart Zoo Gives Perfect Explanation for Why It No Longer Has Elephants." *The Dodo*, April 6, 2016.

211 In 2018, India's Supreme Court: "Order against Caging of Birds Upsets Poultry Farmers in India." *The Poultry Site*, October 31, 2018. See also Saha, Purbita, "Do Birds Have an Inherent Right to Fly?" *Audubon*, April 2016. See also Mathur, Aneesha, and Satish Jha, "Do Birds Have a 'Fundamental Right to Fly'?" *The Indian Express*, December 15, 2015.

211 in 2019, two beluga whales: Mountain, Michael, "Sea Life Trust Is Building the World's First Beluga Sanctuary." *The Whale Sanctuary Project*, August 30, 2018.

212 They hoped to: Khan, Shehab, "Pet Translator Devices Could Let Us Talk to Dogs within 10 Years, Amazon-Backed Report Says." *Independent*, July 22, 2017.

Chapter 8: Melting Ice, Bursting Levees

220 Famed climate scientist James Hansen: Hansen, James, "Disastrous Sea Level Rise Is an Issue for Today's Public—Not Next Millennium's." *Huffington Post*, December 6, 2017.

231 70 percent of the plastic found: Speksnijder, Cor, "Onderzoekers: plasticsoep in Stille Oceaan komt vooral van visserij en scheepvaart." *De Volkskrant*, March 22, 2018.

231 even the most sustainably produced: "Deliciously Ella: The Podcast." *PodBean*. See also "Why a Vegan Diet Is the Single Biggest Positive Change You Can Make for the Planet, with Joseph Poore at Oxford University." *Deliciously Ella: The Podcast*, October 9, 2018.

232 All of a sudden, the papers: Kilvert, Nick, "Would you go vegan to save the planet? Researchers say it might be our best option." *ABC News*, May 31, 2018.

232 "a global transition": Springmann, M. et al., "Options for Keeping the Food System within Environmental Limits." *Nature*, October 2018.

234 In 2014, Americans ate on average: "Schwarzenegger moet zorgen dat Chinezen minder vlees eten." *NOS*, July 25, 2016.

Epilogue: The Beginning of the End

238 developed two space probes: Joosten, Peter, "Sterrenkunde, Supernova's & Ruimtevaart. Met Ans Hekkenberg." *Biohacking Impact*.

242 Women who demanded: Koops, Enne, "Bijzondere vrouwen in de Eerste Wereldoorlog." *Historiek*, May 21, 2015.

Bibliography

Below you will find a list of the books, interviews, and other sources I used while writing this book, arranged by chapter and in order of how they appear in the text. Additionally, where I have used word-for-word quotes, I have included the source from which I took the quote and the page where I have used it. In order to save paper, I have also put together a list of additional notes that can only be found on the website www.roannevanvoorst.com/onceweateanimals. This includes extra information and comments that I felt would be suitable or useful for readers. The website also includes a list of frequently asked, difficult, and important questions about the theme of the book.

Introduction: Inventing a New Color

Adams, Carol J. *The Sexual Politics of Meat: A Feminist-Vegetarian Critical Theory.* New York: Continuum, 1990.

Foer, Jonathan Safran. *Dieren Eten*. Amsterdam: Ambo|Anthos, 2009.

Klein, Naomi. *This Changes Everything: Capitalism vs. the Climate*. New York: Simon & Schuster, 2014.

Monbiot, George. *Feral: Rewilding the Land, Sea and Human Life*. London: Penguin Books, 2013.

Orrell, David. *The Future of Everything: The Science of Prediction*. New York: Perseus Books Group, 2007.

Patterson, Charles. *Eternal Treblinka: Our Treatment of Animals and the Holocaust*. New York: Lantern Books, 2004.

Singer, Peter. *Animal Liberation: A Personal View. Writings on an Ethical Life*. London: Fourth Estate, 2001.

Singer, Peter. *Animal Rights and Human Obligations*. London: Pearson Education, 1976.

Waal, Frans de. *Are We Smart Enough to Know How Smart Animals Are?* New York: W.W. Norton, 2016.

1. How Farmers Can Change the World

My interview with Gustaf took place on May 30, 2018; my interview with Jay and Katja took place on June 14, 2018; I found the information about and interviews with the other farmers online (you can find the links on the list of extras at www.roannevanvoorst.com/onceweateanimals). I also got a lot of inspiration and knowledge from a visit to cattle farmer and cheesemaker Jan Dirk Remeker, which took place on July 10, 2018, and an interview with vegan and farmer's daughter Marloes Boere on August 24, 2018.

Arendt, Hannah. *De banaliteit van het kwaad*. Een reportage. Amsterdam: Moussault, 1969.

Cohen, J. A., A. P. Mannarino, and E. Deblinger. *Behandeling van Trauma bij Kinderen en Adolescenten met de Methode: Traumagerichte Cognitieve Gedragstherapie*. Houten: Bohn Stafleuvan Loghum, 2008.

Coles, Robert, and Erik H. Erikson. *The Growth of His Work*. Boston: Little Brown, 1970.

Erikson, Erik H. *Childhood and Society*. New York: Norton, 1950.

Erikson, Erik H. *Dialogue with Erik Erikson*. Evans, R. I., ed. Lanham, MD: Jason Aronson, 1995.

Erikson, Erik H. *Identity: Youth and Crisis*. New York: Norton, 1968.

Erikson, Erik H. *Life History and the Historical Moment*. New York: Norton, 1975.

Friedman, Lawrence J. *Identity's Architect: A Biography of Erik H. Erikson*. New York: Scribner Book Co., 1999.

2. Why Good People Believe in Bad Stories

For this chapter, I interviewed social psychologist Dr. Melanie Joy on June 12, 2018, and writer Tobias Leenaert (former director of EVA, Ethisch Vegetarisch Alternatief, Ethical Vegetarian Alternative, a Belgian nonprofit organization that promotes and informs people about vegetarianism), on June 5, 2018, and author and science journalist Marta Zaraska on June 14, 2018.

Adams, Carol J. *The Sexual Politics of Meat: A Feminist Vegetarian Critical Theory*. New York: Continuum, 1990.

Bakker, Tom, Willy Baltussen, and Bart Doorneweert. "Concurrentiemonitor blank kalfsvlees." LEI-rapport 2012–025, January 2012.

Beckoff, Marc, and Jessica Pierce. *Animals' Agenda: Freedom, Compassion, and Coexistence in the Human Age*. Boston: Beacon Press, 2017.

Bibliography

Boomkens, René. *Erfenissen van de Verlichting: Basisboek Cultuurfilosofie*. Amsterdam: Boom Uitgevers, 2011.

Boon, Floor. "Vlees zonder bloedvergieten." *Folia*, November 12, 2010.

Bradshaw, Peter. "The End of Line." *The Guardian*, June 12, 2009.

"By-catch." *Radar*, March 8, 2010.

"De verschillen tussen kooi, bio, scharrel en vrije uitloopkip." *Trouw*, February 27, 2013. See also "'Plofkip, gewone kip, biologische kip, scharrelkip.' Meer weten over eten.

Delahoyde, Michael, and Susan C. Despenich. "Creating Meat Eaters: The Child as Advertising Target." *The Journal of Popular Culture*, 1994.

Ercina, A. Ertug, Maite M. Aldaya, and Arjen Y. Hoekstra. "The Water Footprint of Soy Milk and Soy Burger and Equivalent Animal Products." *Ecological Indicators* 18 (2011): 392–402.

"'Gezondheid van Runderen.'" *Levende Have*, 2017.

"Global Meat Production Since 1990." *Rapport Statistica*, November 22, 2018.

Gombrich, E. H. *A Little History of the World*. New Haven: Yale Univ. Press, 2005.

Harari, Yuval Noah. *Sapiens: A Brief History of Humankind*. New York: HarperCollins, 2015.

Hazard, Paul. *The Crisis of the European Mind: 1680–1715*. New York: New York Review Books, 1935.

Israel, Jonathan. *Democratic Enlightenment: Philosophy, Revolution, and Human Rights, 1750–1790*. Oxford: Oxford Univ. Press, 2011.

Keulemans, Maarten. "Geheimzinnig aapachtig oerwezen waggelde tussen onze voorouders." *De Volkskrant*, May 9, 2017.

"Klein brein, maar wel slim." *Trouw*, May 15, 2018.

Klinckhamers, Pavel. "Industriële visserij bedreigt onze Noordzee." *De Volkskrant*, July 5, 2015.

"Livestocks' Long Shadow: Environmental Issues and Options." *FOA*, 2006.

"Meat and Animal Feed." *Global Agriculture*, 2018.

Simon, David Robinson. *Meatonomics: How the Rigged Economics of Meat and Dairy Make You Consume Too Much—and How to Eat Better, Live Longer, and Spend Smarter*. Newburyport, MA: Conari Press, 2013.

Thieme, Marianne, et al. "Dierenactivist is democraat en geen terrorist." *NRC Handelsblad*, April 23, 2008.

Vialles, Noilie. *Animal to Edible*. Cambridge: Cambridge Univ. Press, 1994.

"Wilders wil aparte wet tegen 'dierenterroristen.'" *Trouw*, August 21, 2017.

Winders, Bill, and David Nibert. "Consuming the Surplus: Expanding 'Meat' Consumption and Animal Oppression." *International Journal of Sociology and Social Policy* 24, no. 9 (2004): 76–96.

"World Meat Production 1960–Present." *Beef2live Report*, November 18, 2018.

Bibliography

Intermezzo: We Didn't Know

For this chapter I interviewed historian and lecturer Willem van Schendel on September 4, 2018, and Max Elder, futurologist at the Institute for the Future and research manager at the Food Futures Lab, on July 5, 2018.

Harari, Yuval Noah. *Homo Deus: Een kleine geschiedenis van de toekomst*. Amsterdam: Thomas Rap, 2017.

Joosten, Peter. *Biohacking, de toekomst van de maakbare mens*. 2018.

Nowak, Peter. *Humans 3.0. The Upgrading of the Species*. London: Penguin Ltd., 2015.

Pearson, Ian. *You Tomorrow: The Future of Humanity, Gender, Everyday Life, Careers, Belongings, and Surroundings*. Scotts Valley: Createspace Independent Publishing Platform, 2013.

"#Wereldzonderwerk: dit zijn de banen van de toekomst." *NOS*, March 18, 2017.

3. From Pasty and Peeved to Sexy as Fuck

For this chapter I interviewed Derek Sarno, chef and cofounder of Wicked Healthy, on June 25, 2018, and "Fat Gay Vegan" on September 24 and 25, 2018.

Andrews, Travis M. "Woman Trying to Prove 'Vegans Can Do Anything' Among Four Dead on Mount Everest." *The Washington Post*, May 23, 2016.

Barendregt, Bart, and Rivke Jaffe, ed. *Green Consumption: The Global Rise of Eco-Chic*. London: Bloomsbury, 2014.

Birchall, Guy. "Vegan Mountain Climber Dies on Mount Everest During Mission to Prove Vegans Are Capable of Extreme Physical Challenges." *The Sun*, May 23, 2016.

Brewer, Marilynn, and Wendi Gardner. "Who Is This 'We'? Levels of Collective Identity and Self Representations." *Journal of Personality and Social Psychology* 71, no. 1 (1996): 83–93.

Cherry, Elizabeth. "Veganism as a Cultural Movement: A Relational Approach." *Social Movement Studies* 5, no. 2 (2006): 155–170.

Dell'Amore, Christine. "Species Extinction Happening 1,000 Times Faster Because of Humans?" *National Geographic*, May 30, 2014.

Eggeraat, Amarens. "Waarom haten we veganisten zo?" *Vrij Nederland*, May 28, 2016.

Graham, Sylvester. *A lecture on epidemic diseases generally: and particularly the spasmodic cholera: delivered in the city of New York, March 1832, and repeated June, 1832[sic] and in Albany, July 4, 1832, and in New York, June 1833: with an appendix containing several testimonials, rules of the Graham boarding house*. New York: Mahlon Day, 1833.

Haegens, Koen. "Het Paradijs was Fruitarisch." *De Groene Amsterdammer*, 2007.

Hamad, Ruby. "Why Hitler Wasn't a Vegetarian and the Aryan Vegan Diet Isn't What It Seems." *SBS*, December 14, 2017.

Jasper, J. *The Art of Moral Protest: Culture, Biography, and Creativity in Social Movements*. Chicago: Univ. of Chicago Press, 1997.

Jones, Josh. "How Leo Tolstoy Became a Vegetarian and Jumpstarted the Vegetarian & Humanitarian Movements in the 19th Century." *Open Culture*, December 26, 2016.

Lowbridge, Caroline. "Veganism: How a Maligned Movement Went Mainstream." *BBC News*, December 30, 2017.

Macias, Elena, and Amanda Holodny. "The Eccentric Eating Habits of 9 Ruthless Dictators." *Independent*, November 16, 2016.

Melucci, Alberto. "An End to Social Movements? Introductory Paper to the sessions on New Movements and Change in Organizational Forms." *Social Science Information* 23, no. 4/5 (1984): 819–835.

Melucci, Alberto. "The Process of Collective Identity." In *Social Movements and Culture*, edited by H. Johnson and B. Klandermans, 41–63. Minneapolis: Univ. of Minnesota Press, 1995.

"Miley Cyrus Gets a Tattoo to Show She Is a Vegan for Life." *Live Kindly*, July 10, 2017.

Petter, Olivia. "The Surprising Reason Why Veganism Is Now Mainstream." *Independent*, April 10, 2018.

Poppy, Carrie. "Myth Check: Was Hitler a Vegetarian?" *Skeptical Inquirer*, November 2, 2016.

"Steeds meer mensen gaan voor veganistisch." *Kassa*, December 1, 2014.

Vegan Society. "About Us." 2006.

Viegas, Jen. "Humans Caused 322 Animal Extinctions in Past 500 Years." *Seeker*, April 24, 2014.

4. Giraffes for the Rich, Vegetables for the Poor, and Milk for All

For this chapter I interviewed historian Manon Henzen from www.eetverleden.nl on July 23, 2018. I also learned a lot.

Alblas, Jasper. "Is melk gezond? De feiten en fabels over melk." *Dokterdokter*, December 16, 2016.

Bomkamp, Samantha. "Why Kale Is Everywhere: How Food Trends Are Born." *Chicago Tribune*, September 20, 2017.

Friedrich, Bruce. "Market Forces and Food Technology Will Save the World." TEDx, January 30, 2018.

"Grotere melkveebedrijven en meer melk." *CBS*, May 2, 2017.

Heijmerikx, Anton G. M. "Eten en drinken in de Middeleeuwen." *Heijmerikx*, July 2, 2009.

"Holsteinkoe rendabeler dan Jersey." *Groen Kennisnet*, April 10, 2017.

"Kan Jersey zich meten met Holstein?" *Boerenbond*, January 13, 2017.

Louwerens, Tessa. "Ontmoet de oud-Hollandse koeien." *Resource*, April 20, 2017.
"Meer melk met minder koeien." *The Daily Milk*, May 5, 2017.
"Melkproductie Nederland naar recordniveau." *Nieuwe Oogst*, May 2, 2017.
"Olympische Spelen in de Oudheid." *IsGeschiedenis*.
Reijnders, Lucas, Anne Beukema, and Rob Sijmons. *Voedsel in Nederland: gezondheid, bedrog en vergif.* Amsterdam: Van Gennep, 1975.
Schepens, Juul. "Becel speelt in op de veganistische trend met plantaardige melk." *Adformatie*, August 10, 2018.
Tetrick, Josh. "The Future of Food." TEDx, June 22, 2013.

5. Wanted: Man (20–40), Sporty, Sexy, Vegan

For this chapter I interviewed sociologist and professor Anne DeLessio-Parson on September 17, 2018, and Sean O'Callaghan, better known as @fatgayvegan, on September 26, 2018.

Adams, Carol. *Burger. Object Lessons Series.* New York: Bloomsbury Academic, 2016.
Airaksinen, Toni. "Eating Meat Perpetuates 'Hegemonic Masculinity,' Prof Says." *Campus Reform*, December 4, 2017.
Brighten, Karine. "5 Dating Tips from a Vegan Dating Expert." *VegNews*, February 2, 2017.
Brubach, Holly. "Real Men Eat Meat." *New York Times*, March 9, 2008.
"Dit is waarom mannen van vlees houden." *nu.nl*, November 20, 2016.
Fox, Maggie. "Why Real Men Eat Meat: It Makes Them Feel Manly." *NBC News*, November 21, 2012.
Gander, Kashmira. "Vegan Dating: The Struggle to Find Love When You've Ditched Steak and Cheese." *Independent*, February 22, 2017.
Kucan, Daniel. "You Eat Like a Girl: Why the Masculine Dilemma Towards Veganism Is No Dilemma at All." *Huffington Post*, August 19, 2013.
Leeuwen, Louise van. "Over Hoe de Veganist en de Niet-Veganist nog Lang en Gelukkig Leefden." *Eigenwijs Blij*, September 7, 2016.
Dr. Lockwood, Alex. "Why Aren't More Men Vegan? It's a Simple Question—with a Complicated Answer." *Plant Based News*, February 21, 2018.
"Looking for Love? Here's the Official Top 4 Vegan Dating Websites." *The Plantway*, https://www.theplantway.com/best-vegan-dating-websites/.
May, Zoe. "What Happened When I Tried to Meet Guys Using Vegan Dating Apps." *Metro*, May 9, 2018.
Mycek, Mari Kate. "Meatless Meals and Masculinity: How Veg* Men Explain Their Plant-Based Diets." *Food and Foodways* 26, no. 3 (2018): 223–245.
Rozin, Paul, Julia M. Hormes, Myles S. Faith, and Brian Wansink. "Is Meat Male? A Quantitative Multimethod Framework to Establish Metaphoric Relationships." *Journal of Consumer Research*, 2003.

Starostineskaya, Anna. "Veg Speed Dating Debuts in 20 Cities in February." *VegNews*, January 11, 2017.

Stibbe, Arran. "Health and the Social Construction of Masculinity in Men's Health Magazine." *Man and Masculinities*, July 1, 2004.

Vadas, Skye. "Inside the World of 'Vegansexualism'—the Vegans Who Only Date Other Vegans." *Vice*, October 10, 2016.

"Vegansexuality Explained." *Happy Cow*.

Walker, Jennyfer J. "I Tried to Find Love on Vegan Dating Apps." *Vice*, January 25, 2018.

Winsor, Ben. "'Vegansexual' Is a Thing and There's More Than One Reason Why." *SBS*, June 17, 2016.

6. Plant Overdose

I was advised on this chapter by dietitian and nutritionist Saraï Pannenkoek. I also received helpful comments on a number of questions from Caleb Finch, professor of the neurobiology of aging at University of Southern California.

Atchley, R. A., D. L. Strayer, and P. Atchley. "Creativity in the Wild: Improving Creative Reasoning Through Immersion in Natural Settings." *De Fockert* 7, no. 12 (2012). See also Bakker, Shannon. "NUcheckt: Vitamine B12-tekort komt waarschijnlijk minder voor dan beweerd." *NU.NL*, September 18, 2018. See also Bluejay, Michael. "Humans Are Naturally Plant-Eaters According to the Best Evidence: Our Bodies." *Michael Bluejay*, June 2002, updated December 2015.

Bratman, G. N., J. P. Hamilton, K. S. Hahn, et al. "Nature Experience Reduces Rumination and Subgenual Prefrontal Cortex activation." *Proceedings of the National Academy of Sciences of the United States of America* 112, no. 28 (2015): 8567–8572.

"Duits onderzoek: veganistische producten vaak vet en ongezond." *Trouw*, April 4, 2014.

Katan, Martijn. *Wat is nu Gezond*. Amsterdam: Bert Bakker, 2017.

Schüpbach, R., R. Wegmüller, C. Berguerand, M. Bui, and I. Herter-Aeberli. "Micronutrient Status and Intake in Omnivores, Vegetarians and Vegans in Switzerland." *European Journal of Nutrition* 56, no. 1 (2017): 283–293.

Intermezzo: A School Trip to the Slaughterhouse

Grunberg, Arnon. "We slachten hier 650 varkens per uur." *NRC Handelsblad*, August 23, 2016.

7. It's the Law, Stupid!

For this chapter I interviewed Edwin van Wolferen, communication advisor at the Rijksdienst voor Ondernemend Nederland (Netherlands Enterprise Agency, RVO.nl) and connected to the Central Authority for Scientific Procedures on

Animals and the Netherlands National Committee for the protection of animals used for scientific purposes, on October 25, 2018. I also interviewed Lauren Choplin from the Nonhuman Rights Project on August 16, 2018, and Steven M. Wise, from the same organization, on September 11, 2018. On January 15, 2019, I interviewed Carol Saxon from the Whale Sanctuary. I also got a lot of advice and comments on this chapter from Noor Evertsen, consultant and researcher at Dier & Recht.

Bekoff, Marc. *The Emotional Lives of Animals: A Leading Scientist Explores Animal Joy, Sorrow, and Empathy and Why They Matter.* Novato, CA: New World Library, 2007.

Berns, Gregory. "Dogs Are People, Too." *New York Times*, October 5, 2013. See also Berns, Gregory. *What Is It Like to Be a Dog.* New York: Basic Books, 2017.

"Dolfinarium." *RAMBAM*, March 2, 2016. See also Riepema, Sanne. "'Circus' Dolfinarium onder vuur na uitzending Rambam." *AD*, May 9, 2016. See also Gold, Michael. "Is Happy the Elephant Lonely? Free Her, the Bronx Zoo Is Urged." *New York Times*, October 3, 2018.

"Human Persons." *Small Change*, August 3, 2017.

Janssen, C. "Leve het Dier." *De Volkskrant*, October 3, 2015.

Meijer, Eva. *Dierentalen*. Amsterdam: Coetzee, 2016.

Pasha-Robinson, Lucy. "Hundreds of Animal Species 'Being Consumed to Extinction'." *Independent*, October 19, 2016.

Plotnik, Joshua M., Frans B. M. de Waal, and Diana Reiss. "Self-Recognition in an Asian Elephant." *PNAS* 103, no. 45 (2006).

WWF. *Living Planet Report*. WWF Global, 2014.

8. Melting Ice, Bursting Levees

For this chapter I interviewed Henk Westhoek, Food and Agriculture program manager at the PBL (Netherlands Environmental Assessment Agency), on August 22, 2018, and George Monbiot, science journalist and author, on November 15, 2018.

Boer, Imke de. "Mansholt Lecture: Circular Agriculture, a Good Idea?" *WURtalk 30*, November 1, 2018.

"De Vloek van het Vlees: slecht voor klimaat, milieu, en mensheid." *NRC Handelsblad*, December 21, 2018.

Kamsma, M. "Wat als we stoppen met vlees eten?" *NRC Handelsblad*, October 25, 2018.

Poore, J., and T. Nemecek. "Reducing Food's Environmental Impacts Through Producers and Consumers." *Science* 360, no. 6392 (2018): 987–992.

Poore, J. "Back to the Wild: How Nature Is Reclaiming Farmland." *New Scientist* 235, no. 3138 (2017): 26–29.

Springmann, Marco, H. Charles, J. Godfraya, M. Raynera, and Peter Scarborough. "Analysis and Valuation of the Health and Climate Change Cobenefits of Dietary Change." *PNAS*, 2016.

Stehfest, Elke, Lex Bouwman, Detlef P. Van Vuuren, Michel G. J. Den Elzen, Bas Eickhout, and Pavel Kabat. "Climate Benefits of Changing Diet." *Climatic Change* 95, no. 1–2 (2009): 83–102.

Westhoek, Henk, Jan Peter Lesschen, Adrian Leip, Trudy Rood, Susanne Wagner, Alessandra De Marco, Donal Murphy-Bokern, Christian Pallière, Clare M. Howard, Oenema, Oene, and Mark A. Sutton. *Nitrogen on the Table: The Influence of Food Choices on Nitrogen Emissions and the European Environment. European Nitrogen Assessment Special Report on Nitrogen and Food.* Edinburgh: Centre for Ecology & Hydrology, 2015.

Westhoek, Henk, Jan Peter Lesschen, Trudy Rood, Susanne Wagner, Alessandra De Marco, Donal Murphy-Bokern, Adrian Leip, Hans van Grinsven, Mark A. Sutton, and Oene Oenema. "Food Choices, Health and Environment: Effects of Cutting Europe's Meat and Dairy Intake." *Global Environmental Change* 26 (2014): 196–205.

Westhoek, Henk, Trudy Rood, Maurits van den Berg, Durk Nijdam, Melchert Reudink, Elke Stehfest, and Jan Janse. "The Protein Puzzle." *PBL*, 2011.

Zanten, Hannah van, Mario Herrero, Ollie Van Hal, Elin Röös, Adrian Muller, Tara Garnett, Pierre J. Gerber, Christian Schader, and Imke J.M. de Boer. "Defining a Land Boundary for Sustainable Livestock Consumption. RESEARCH REVIEW." *Global Change Biology* 24 (2018): 4185–4194.

Epilogue: The Beginning of the End

For this chapter I interviewed Lauren Ornelas, founder and CEO of the Food Empowerment Project, on September 25, 2018, and "green protein commissioner" and policy advisor Jeroen Willemsen on August 10, 2018.

Monbiot, George. *How Did We Get into This Mess? Politics, Equality, Nature.* London: Verso, 2016.

Weisman, Alan. *De wereld zonder ons.* Amsterdam: Atlas, 2007.